Permaculture
PLANTS

Agaves and Cacti

Jeff Nugent

Sustainable Agriculture Research Institute

www.permacultureplants.net

Sustainable Agriculture Research Institute
PO Box 10, Nannup
WA 6275, Australia
Telephone (08) 975 61271

Second Edition September 2006

National Library of Australia
Cataloguing-in-Publication
Nugent, Jeff, 1953-
Permaculture plants: agaves and cacti

Includes index
ISBN 0 9586367 0 2.
1. Permaculture. 2. Plants Useful.
3. Ethnobotany. 4. AGAVACEAE. 5. CACTACEAE
Sustainable Agriculture Research Institute
Title

631.58

Front cover: *Lophophora williamsii* (peyote) in foreground prickly pear and *Yucca brevifolia* in background with *Agave americana* leaf wrapping around to form a background on spine and back cover.
Back cover: Left: *Cereus hildmannianus* supporting a *Hylocereus sp.* in Kenya. **Right** (from top): *Hylocereus trigonus* in fruit; *Opuntia robusta* in fruit; Commercially marketed dragon fruit; *Pereskia aculeata* in flower.
Drawing previous page: *Lophophora williamsii* (peyote).
Some cacti store water in their root system, rather than the more obvious above ground store. The plant apparently does not require spines to protect its stores which are safe below the ground. In dry conditions the plant has the ability to pull itself into the soil. This allows it to reduce exposure to the drying conditions. The cacti with this ability are known as "living roots". When the people of Mexico dug up this plant, the similarity of the plant to a spirit figure was supported by the plants psychedelic effects. The spirit Mescalito was at the heart of the desert dwellers religious ceremonies and has been in use for more than four thousand years.

Sustainable Agriculture Research Institute (SARI) www.permacultureplants.net

SARI is based in Nannup, Western Australia. Extension services available through SARI include: property selection and design \ nursery stock supply \ and planting of systems. SARI runs courses on Permaculture including the Permaculture Design Certificate Course. This is SARI's second major publication.
The author works as a researcher, a designer and an educator with SARI.

Also Available from SARI

Permaculture Plants: A Selection
Permaculture Design Course - Bill Mollison
(44 hours of audio cassette tapes available soon as mp3)

Soon to be Published

Permaculture Plants: Palms and Ferns
Permaculture Plants: a Tropical Companion
Aquaculture, the Permaculture Perspective by Bill Mollison
(Audio MP3 set)

Other Services on Offer

Permaculture Certificate Courses - At SARI / or by correspondence.
Courses can be tailored to suit special interest groups.
Permaculture Consultancy

Dedication

This series of books is dedicated to future generations who face the enormous task of repairing a squandered environment with diminished resources.

In particular, in a time when one fifth of the world's human population dies from starvation due to desertification, this book is dedicated to those repairing the deserts of human making.

Thanks

Thanks must firstly go to my family who put up with my mood swings, vacant stares and neglect of duties in the process of putting this book together. Especially thanks to my wife Jill who stood in for me on many occasions.

Tanya McAllan and Julia Boniface thank you for doing the proof reading.

Finally the ultimate thanks to Bill Mollison who has done so much toward making this planet a better place to live.

Contents

I planted these agaves across a badly eroded stock path in the village of Ngare Ndare in Kenya. The previous path took just 3 years to become the massive erosion gully seen on the right I used basal suckers from a few sisal plants growing on the site. It took many hours to break up the compacted earth sufficiently to plant. All plantings were made into shallow trenches cut on contour. This served to sink the occasional, but heavy, rains into the ground rather than promoting gully erosion.

Two and a half years later they had successfully rehabilitated the site.

Preface

It is with humility that I present this book. The driving force to present it is knowing that it will help other designers to become more effective in their work.

Although I have made every effort to ensure that the information in this book is as accurate as possible, most of this information has been gleaned from references. Many of these are very old and I acknowledge that there may be some information that is incorrect. I take no responsibility for how people use this information or for any wrong information in the book. I hope to get feed-back from people so that subsequent editions will become more useable. Much of the book is essentially written in broken English. I have tried to avoid the use of unnecessary verbiage so that the book is more useable to people who do not have English as their first language. I trust the English speaking reader will bear with this.

I never actually set out to write this book it just fell out of the pages of <u>Permaculture Plants: a tropical companion</u> (a volume in progress). It seemed logical to publish it as a separate volume in much the same way as I am doing with <u>Permaculture Plants: palms and ferns.</u>

I have assembled all of the uses for the families AGAVACEAE and CACTACEAE that I could find. Added to that is any information I could on the plants habitat, climatic requirements and cultivation tips. I am constantly amazed that our "civilisation" has never taken the time to do this across all families of plants.

Certainly one of my greatest obstacles has been in naming species. As a rule of thumb I have listed every botanic name assigned to a species (the most recently accepted one is first and in bold print) and also every common name I have come across (with the exception of "that prickly little bastard"). I have witnessed first hand the value of having all names available for location of seed and to ensure that two parties are speaking of the same species.

I am not a taxonomist so I have relied on others to sort out species names. As there are so many different ideas as to how the species should be classified it has been necessary to adopt a standard. Rightly or wrongly, with the cacti naming, the final word has been Hunt (1992). This is the list that most suppliers will be working from with CITES (See Glossary) listing. Most of the names fit this reference although there are some species mentioned in older texts which I have not been able to match. When in doubt I have listed them under the name given in the original text.

It should be noted that the family AGAVACEAE is often divided into various other families:

AGAVACEAE includes Agave spp. and Yucca spp.)

DRACAENACEAE includes *Dracaena spp.* and *Sansevieria spp.*

DRACAENACEA have also been classed in the family ASPARAGACEAE

PHORMIACEAE to include *Phormium spp.*

This genus has also been classed as XANTHORRHOEACEAE

ASTELIACEAE to include *Cordyline spp.*

NONLINACEAE to include *Dasylirion spp.*

For the purposes of this book they have all been classed as AGAVACEAE.

INTRODUCTION

The cacti are a large family exclusively native to the Americas. Exclusively that is except for the mistletoe cactus *Rhipsalis baccifera*. This cactus, with its edible fruit, is native to a large area of tropical and sub tropical America, but is also native to Sri Lanka, West Africa and Madagascar.

It almost certainly originated in the Americas, but exactly how this plant ever reached places like the remote island of Madagascar, in pre-Columbian times is a mystery which may never be resolved. It is likely that migratory birds carried the seed in their intestines. There is also a possibility that humans took them. The Indonesian people who settled Madagascar some 1200 years ago were notorious travellers and traders.

It is equally fascinating to note that the migration of plants by birds is considered natural and that the introduced species is considered indigenous to its colonised area. When humans move a species however, it is considered an exotic and somehow taboo. If it becomes established in it's new region it is regarded as feral. We distinguish between plants moved by animals and humans, yet we are also products of nature.

Consider the movement of a different cactus coming into Madagascar from the Americas. In 1768 Count de Modave, a French colonial, re-founded a French settlement at Port Dauphin on the south east coast of Madagascar, along the trading route to Asia. He introduced a Mexican prickly pear *(Opuntia sp.)* from a botanic garden at Reunion Island.

Modave's purpose was to defend his fort with a barrier of prickly pear. The species he chose was heavily armed with 100mm long spines and was capable of forming impenetrable barriers 4m tall. Modave's paranoia was not unfounded, as the people took exception to becoming slaves and finally overthrew the fort.

The prickly pear continued to grow and to fruit and by 1900 when the French returned it was a major part of the local ecology and the foundation of the local economy. The local people (Antandroy) called it *raketa* after the tennis racket shaped leaves. They had planted *raketa* hedges to contain their zebu and to defend them from cattle thieves.

The purple fruit was an important part of their diet and the leaves were a source of drinking water during droughts. The Antandroy found that they could burn the thorns off the *raketa* and then their cattle, the zebu would eat them, supplying them valuable food and water. *Longo Tandroy sy raketa* was an Antandroy proverb of the time -"The Antandroy and the prickly pear are one".

A new French assault which began in 1900 was impaired by the *raketa*. The whole landscape was like a giant labyrinth of thorny barriers. Every movement was monitored by the Antandroy and the narrow paths that the French took between hedges of *raketa* would suddenly be blocked by a heap of *raketa* pads. Before the French could deal with the *raketa* barrier they would fall victims to a rain of spears from the other side of the hedges.

In 1925 the French introduced the cochineal beetle (actually a woolly aphid) *(Dactylopius opuntia, syn. Dactylopius cochenillifera)* to the district as a way of breaking the Antandroy defence. By 1928 the cochineal beetle had a strong-hold and within a couple of years *raketa* had all but disappeared from the landscape.

The ensuing famine of 1931 saw half the population die or move. Many of the children were traded to the east coast by their parents in exchange for corn. The Antandroy-*raketa* alliance was broken and the French took control.

There are many morals to this story, but surely an important one is not to base our existence on too few species. Had the Antandroy relied on half a dozen other hedge species of the families AGAVACEAE and CACTACEAE and mixed with that some members of the EUPHORBIACEAE family they would probably have retained their lands. History can be such a random set of events.

Although the family AGAVACEAE is not restricted in natural distribution to the Americas, it is an American agave that has become a major export product of the same island. Madagascar is the fourth largest producer of sisal fibre on the planet.

About this Book

It is not the purpose of this book to cover the vast topic of Permaculture. Rather it is a resource list that I hope will help Permaculture practitioners with their efforts. I feel sure that the title will pull in some outsiders so I have included in this book a quick overview of Permaculture so that the novice can see the big picture.

In the late 70's and early 80's the species list in "Permaculture 1" by Bill Mollison and David Holmgren (Tagari Publications), was all that we had as easily accessible tools in our design work. The rest was a matter of wading through volumes of material and wishing it was already compiled. Little has changed in the late 90's except there are more texts available to wade through.

Permaculture Plants: A Selection was an effort to help to reduce the wading, allowing more time for planning and planting. This volume is an extension of that process, concentrating on two very important families of plants AGAVACEAE and CACTACEAE. Both families were barely covered in Permaculture Plants: A Selection, yet they have been major contributors to the well-being of many of the indigenous people of the Americas. Evidence suggests that they have been part of the human diet for more than 9000 years.

Today in the western world, most small-scale growers of these plants do so for ornamental purposes. There is an enormous amount of energy and resource gone into producing plants with 'improved' flowers or spines. If a fraction of this energy had gone into developing the fruits of the various edible species then I feel confident that cacti fruit would be as common in the markets as apples, oranges, coconuts and pineapples.

But sadly this hasn't happened. Instead, we see species being threatened with extinction because of human folly. We need to try to preserve these species but should also be very careful not to contribute to the problem by purchasing seed or plants from collectors who deplete wild resources. The global community is working to protect these endangered species from indiscriminate collectors by establishing the CITES check list. CITES certificates are required for importation of seeds listed in the check list into any signatory country. All cacti are CITES listed.

I am sickened to hear stories of rare cacti, many hundreds of years old, being removed from the wild and placed into private collections. Please purchase with awareness.

At no time in the history of the planet has there been such a great opportunity to shift species around the planet. International flight means that even those seeds which lose viability very quickly can be transported. There is of course a need for some care in selecting which species do get introduced. It is important to monitor their progress before widespread introduction to an area, but ultimately most species are complete unknowns in a different environment until they are trialed there. Certain species are prohibited imports into some countries because of their status as rampant weeds. There are examples, particularly with the introduction of species of prickly pear cacti *(Opuntia spp.)*, which have become rampant weeds. In trialing species in new environments a certain amount of care must be exercised, but this should not be an excuse for doing nothing. We are running out of time.

There is a major global move toward prohibiting the importation of species. This movement seems to be prompted by multi-national companies who perceive that Permaculture could interfere with their command over the planet. That command is now all but complete as every government on the planet faces bankruptcy and is selling off the last of their assets to balance the books.

This movement to suppress species shift is supported by a sector of the environmental movement. The catch cry is "invasive species", yet most invasive species (perhaps with the exception of *Homo sapiens*) are only growing in badly damaged ecosystems and are usually doing a valuable job of holding the soil together and in the case of the feral prickly pear, it was about the only plant that cattle would not eat, so it had an unnatural advantage in over-grazed country. Something always seems to step into an ecosystem to solve problems, even overgrazing cattle.

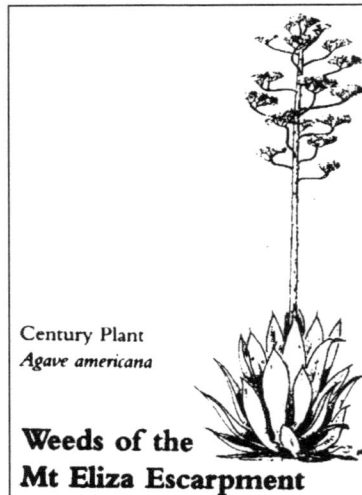

Century Plant
Agave americana

**Weeds of the
Mt Eliza Escarpment**

*Part of a sign in Perth, Western Australia.
On closer inspection the century plant is doing
most of the work in holding the hill together.*

Ecosystems are not fixed, but dynamic. It is fortunate that we have the diversity of species from around the world to get us through these times of rapid change, loss of species and decline of ecosystems.

We are losing species on this planet faster than they are being catalogued, as the industrial world exploits the last corners of the planet and the last wilderness is lost. We are also losing the information that indigenous peoples had on how these species could be utilised. Perhaps the only chance of saving many of the species is to grow them in our own systems. Indeed many of our domesticated fruits are now unknown in the wild.

This book is really about conservation - conservation of soil, fossil fuel, human energy, species and genotypes of species. It is also about conservation of knowledge. Ultimately it is about conservation of the human species which is obviously doomed unless we change our priorities and shift our direction towards well designed sustainable systems.

The plants listed in this book have great potential to turn the problems of food scarcity and desertification into solutions of food security and abundance.

As global warming and climatic destabilisation become obvious and dominant forces, we can brace ourselves against drought conditions. As this book goes to press the ground-water in our district is lower than ever before in human memory. Changes are happening. The plants in this book are adapted to uncertain rainfall.

These are the tools - now use them.

OVERVIEW OF PERMACULTURE

The best way of using the tools is by designing them into the system using the principles of Permaculture. It is strongly recommended that those unfamiliar with Permaculture seek out and study Introduction to Permaculture and Permaculture: A Designers Manual, both by Bill Mollison (Tagari Publications).

Permaculture Species

Compare the array of plants and animals available to us, with contemporary, industrial agriculture, where single species are grown to the absolute exclusion of anything else and over vast expanses of land.

There are about 15,000 plants recorded in the literature as edible. It seems likely that there are many more still unrecorded. There are also the plants that feed us indirectly. Flowers produce nectar from which our bees make honey. Legumes, grasses and trees feed our animals to give us meat, eggs, milk and so on. Other species of plants provide us with medicines, dyes, fibres, timber, fuel, fences, windbreak and shelterbelt. Some plants fix atmospheric nitrogen and others suppress fire.

"The tool with the greatest potential for feeding men and animals, for regenerating the soil, for restoring water systems, for controlling floods and droughts, for creating more benevolent micro-climates and more comfortable and stimulating living conditions for humanity is the tree." - J. Sholto Douglas et al, Forest Farming (Robinson & Watkins Books, 1981)

Design Principles

The greatest criticism of Permaculture the author has ever heard, is that it is "just common sense really". Sadly common sense is not common though. Permaculture is common sense, but the thing that separates it from other methods of providing human needs is design. The design is based on observations we have made of patterns of natural ecosystems and how they function. In our planned ecosystems of useful plants and animals, different plants and animals in the system contribute to the integrity of the whole and all benefit from it. Monoculture with its associated destruction of the soil, its heavy demands for energy and use of chemicals becomes redundant.

There are a few fundamental principles which must be taken into account. Every element in a system should serve many functions. Every function should be supported by many elements. Every action we make should achieve many results.

When designing a property, first consider the **house site(s)**. These are determined by **sun facing slopes** in cooler climates, **gravity fed water** (where possible), **economical access** and **site repair** (with everything else equal, it is better to build a house on a site that is degraded and enhance it than to select a beautiful site and destroy it.

Once the house(s) is placed we can draw several overlays onto a map of the property. The first is **zones**, which can be easily summed up by the rule of thumb **oftenest is nearest**. By this we mean that the things that we visit most often are placed nearest to the house(s). The benefits of this approach to the occupier(s) of the house should be obvious. The **home garden** and grafted fruit trees are near the **centre of activity** where they get the maximum attention and where the harvests are close to the kitchen. The **fruit forests** and **animal forage systems** are placed further away as they do not need visiting very often. Through this whole system can be woven a **multi-purpose walk** which takes in all the daily chores in one passing. Things needing most attention are placed closest to the multi-purpose walk.

The second overlay (**sectors**) looks at blocking, directing and even harvesting natural energies entering the system. Cold winter winds are blocked, cool summer winds are encouraged, hot summer sun can be screened from some elements such as the house and shade loving plants, whilst the winter sun is allowed to penetrate deep into the system.

The interface between two ecosystems such as land and lake, forest and grassland, or ocean and land is always much richer in species than either of the two systems. These edges are habitat to species from both ecosystems but also to a range of species quite specific to that edge. We can use this **edge effect** in our designs so as to maximise the productivity of a system.

In nature we see **stacking** of plants where different species occupy every level of a forest or woodland. Even in arid and semi-arid conditions we see stacking at work. It is possible with careful planning to arrange a stacked ecosystem of plants, all or most of which are in some way benefiting the whole and providing us with products.

In the south west of Western Australia, as with many parts of the world, we are especially aware of increasing ultraviolet radiation levels. The effects on plants are not well known. Early research suggests that mature plants have a better chance of survival. Plants from the highland tropics are already adapted to higher ultraviolet radiation levels.

The effects on human health of increased ultraviolet radiations are well documented and include increases in skin cancer, cataracts on eyes and general breakdown of the immune system. We try to avoid long periods of exposure to the sun by wearing protective clothing. Patterns in our property designs can also help to protect us from too much exposure.

Shaded walks should be incorporated into the system. Roads and pathways are vital to our systems if we are to travel efficiently to different parts. We often find ourselves having to move around outside in the heat of the day, when ultra violet radiations are potentially at their worst.

Starting with the most used walkways establish trellis and large shade trees over these paths and roads. The most frequently used sections of the multi-purpose walk, for example the path to the chook (chicken in America and hen in England) roost, can be trellised with a mixture of deciduous and evergreen vines.

By arranging a balance of deciduous and evergreen shade plants it is possible to provide a warm place to sit on cooler days.

Shaded play can protect children. Keeping children out of the sun is as difficult as keeping a hat on a child. It therefore makes sense to arrange the system so that those places where children are attracted are shaded.

The **adventure playground** should be sited in such a way that it attracts children out of the sun as much as possible. Playground equipment such as trampolines, slides etc. can be spaced along the shaded multipurpose walk. Often these apparatus can be a goal along the walk.

Living fences can reduce construction costs, material usage and minimise maintenance. Many of the agaves and cacti lend themselves to living fences. By planting appropriate species close enough together a hedge can be grown that will not allow stock or wild animals to pass. If smaller animals are a problem, mesh can be incorporated also.

The Selection of Species

Naturally not all species will suit a particular site. Although we are usually engaged in maximising the products and functions of a plant, the primary concern when fitting a species into a system is to ensure that its needs will be met. Each plant needs optimum conditions so as to give optimum production. There are exceptions to this, for example a cork oak growing on dry, rocky outcrops produces a superior cork to one growing in more ideal conditions.

Plants can be placed into micro-climates so as to accommodate their special requirements. These microclimates can optimise aspect, slope and shelter. Soil types are also a consideration. Natural rainfall of a site is not always optimum for a particular species. Irrigation may be necessary for the home garden in dry situations. Alternatively many of the plants listed in this book may need to be placed on well drained slopes so as not to become waterlogged by greater rainfalls than they would normally receive. Many of these survive our Mediterranean climate but the combination of wet and cold handicaps others. There should be some species that will survive most climates.

Given that the global climate is changing, it is reasonable to include species which will survive the current climate but which may excel if the climate shifts in a given direction. The plants in this book should have obvious benefits for this purpose. Since researching this book I have committed myself to planting large numbers of *Agave americana* and to commence a collection of other useful agaves. They do extremely well in this climate (warm Mediterranean) and established plants represent food security beyond the wildest dream. Whole nations of native Americans can not be wrong. In 1993 an estimated 1.2 million people derived their livelihoods from agaves and cacti.

Computer models of global warming suggest that we may lose rainfall in many areas (in the Australian southwest for example as with many other places). This is another good reason to establish AGAVACEAE and CACTACEAE species now.

Rare species can be established and propagated from cuttings or seed and held secure in cultivation. A species is not secure in a region until it is distributed enough that a few destructive events (such as fire or bulldozers) do not eliminate most of the material.

Indoor Plants as Air Filters

Modern Western homes are filled with gases given off from plastics, resins, solvents and glues. Many of these gasses are dangerous to the health of the people inhabiting the houses. Obviously we should try to avoid living in such conditions but often we have little choice.

Work conducted by Wolverton (see References) shows that plants have varying abilities to filter different pollutants from the air and also to adjust air humidity. Of the 50 popular house plants tested 4 were of the family AGAVACEAE and 1 was from the family CACTACEAE (see air filter in index).

It seems likely that all plants tolerant of indoor conditions will have some ability in this area, but a common argument against the use of indoor plants is that they tend to use oxygen at night. AGAVACEAE and CACTACEAE, however are among the plants which have developed a different mechanism. As a means of conserving moisture these plants store gases for exchange until the night when they open their stomates. This process is known as Crassulacean acid metabolism (CAM). Their oxygen release therefore occurs at night making them ideal companions to other indoor plants which release their oxygen during the day.

Accessing Plants

Identifying the best species for a situation is only part of the process. Availability of species also has to be considered.

Many suppliers and also cactus and succulent societies are listed in the appendix.

PROPAGATING PLANTS

Seeds

The cheapest and sometimes the only way of accessing a species is by seed. Plants propagated by seed are generally variable and it may become necessary to plant a lot of seeds and select the best specimens from these to propagate from. This can be done through selective breeding or by vegetative propagation such as cuttings. Grafting of select cacti material is also an option, and is popular amongst those who keep cacti for show.

All species have some mechanism for delaying germination until the seed has dispersed. Some species are simple and others more complex to germinate. I have found most of the AGAVACEAE and CACTACEAE seed that I have attempted to germinate is fairly straight forward.

Here is my preferred method:- The best medium is a sterile peat moss or the more environmentally friendly coconut moss made from coconut fibre. Mix the moss thoroughly with equal parts of sand. This medium should be barely damp. It should not be possible to squeeze any free moisture from a handful. All seed should be free of fruit and other contaminants so as to avoid carrying and/or feeding pathogens.

The problem with small seed such as cacti seed in conventional seed flats is that the medium can easily dry out or become waterlogged from over watering, and in both instances it will not germinate. Sealed plastic containers marketed in Australia with yoghurt or ice cream in them, can be half filled with medium and sown with seed. I find it best to pack the medium firmly first. Another container of the same size works well. Then sprinkle the seed onto the surface, sprinkle a little more medium on to the top and pack lightly once more. The lid is sealed to contain the moisture and the container placed in a warm position where some daylight strikes the container. They are then free to germinate and grow until they reach a manageable size for transplanting to pots or into the ground. Check moisture levels occasionally. If they are too dry sprinkle a little water on. If they are too moist leave the lid off for a while.

Gibberellic acid (GA-3) is available commercially and treatment of cacti seed with the acid is claimed to improve germination considerably. In the case where one is dealing with rare seed it obviously makes sense to use every germination aid possible.

GA-3 can be obtained by post from JL Hudson, Seedsman (see appendix).

The first year is the critical time for seedlings and where these can be raised in nursery situations they are likely to have best survival rates.

Cuttings and Suckers

Most species of cacti are able to be grown from cuttings. Generally a cutting needs a week or two without water to give the wound a chance to callous.

The agaves can be propagated by digging up the root suckers from around the base of adult plants.

Tissue culture is a method of propagation which can establish many clones from a small amount of plant material. Basic laboratory conditions are required but it may be appropriate where a species is very rare or when many plants are required quickly.

AGAVACEAE

Agavus is Latin for "illustrious". Most species have leaves which yield fibre and saponin.

Agave spp.

mescal
century plant

Genus of approximately 300 species \ the same common name is often applied to many species \ generally a desert plant \ able to withstand temperatures to about minus 5^0C \ drought tolerant with transpiration rates among the lowest of all plants \ able to grow in alkaline, poor soils \ not tolerant of waterlogging \ plants live from 5 to 20 years before flowering and dying \ they usually produce many new bulbils from their base before dying \ the bulbils can be removed and planted at good spacing to provide the next crop \ the fermented drink pulque is harvested from many agave species, but usually not until the plant is 8-10 years old \ pulque can be used after one days fermenting as a yeast substitute, by mixing about 1 litre of pulque per 14 cups of flour \ it is usually three years from planting before the plant leaves can be harvested for the first time for its fibre.

In Mexico agaves are said to have some 400 uses.

The Havasupi tribe traded young buds to the Hopi which they weighted with stones so as to deform the buds into balls. They were baked in earth ovens with leaves \ the preparation was soaked in water to produce a drink.

Many agaves occur naturally with *Yucca spp.* and various cacti.

Propagation from seed at 20-30^{0}C with 13-14 hours daylight per day and remove from closed, humid environment a few days after germination.

Pulque

Pulque is a fermented drink made from the juice (*aguamiel*-meaning honey water) of the agave. This can be harvested in two ways. The first is by cutting the leaves off the plant and juicing the heart (see tequila below).

The other method is to cut off the emerging flower stem (the plant only flowers once in a life-time) and create a hollow in the top of the plant. The plant continues to push *aguamiel* to the flower but instead it becomes trapped in the hollow.

A long-necked gourd[1] is modified to harvest the juice. A hole is made in the top and one in the bottom.

---

[1] *Lagenaria vulgaris,* the long necked gourd, also known as bottle gourd and calabash is native to Africa, but has been in use in the Americas for about **12,000 years**. It has been used throughout the world as a vessel for liquids.

The neck is inserted into the pool of *aguamiel* and the hole at the wide end of the gourd is sucked until all of the liquid is held in the gourd.

It is then emptied into a barrel and is ready for fermentation.

Madre pulque (mother of pulque) is prepared by allowing a small amount of *aguamiel* to ferment naturally (usually for 10 days). This acts as a starter yeast and is added to *aguamiel* to induce rapid fermentation. The pulque is ready to drink in a day or two.

The liquid is white in colour and smells of yeast. In fact it is still fermenting. It can even be used as a yeast for bread making. The flavour is apparently acquired, but for newcomers it tastes worse than it smells. Pulque is not bottled and spoils within a few days. The alcoholic content is fairly low (up to 6%) but can be removed altogether by heating for 5 minutes at 80^0C.

Pulque is a favoured drink of Mexicans, who drink it in large quantities. It is rich in many amino acids, vitamins and some sugar and is very cheap. Pulque is often flavoured with fruit including cherry, banana and pineapple.

The use of pulque dates back more than 1700 years. Mayahuel was the Aztec goddess of pulque.

About 30,000 hectares of land was dedicated to pulque production in Mexico in 1993, producing about 400 million litres annually.

Mescal

Mescal or mesquite, as it is sometimes known is essentially distilled pulque, although it can be arrived at using different processes. The drink is called tequila if it is made in the Mexican town of Tequila in the state of Jalsico.

The leaves of the plant are removed.

The hearts (*cabezas*) are removed and split down the middle. They are steam cooked for 36-48 hours and then crushed to remove the juice which is allowed to ferment in vats for a few days. The ferment is double distilled to about 57% alcohol. The first liquid to condense usually contains a high percentage of harmful methanol and is separated from subsequent liquid.

In Mexico in 1993 about 20,000 hectares of mescal agaves (some intercropped with beans and corn) and 18,000 hectares of tequila agaves were under cultivation producing annually some 20 million litres of mescal and 70 million litres of tequila. Some 500,000 plants are also harvested from the wild, placing stress on wild populations. In the same year world demand for tequila outstripped supply and other carbohydrate sources were used to meet the shortfall.

Agave amaniensis
blue sisal

Corrugated leaves \ yields more fibre than *Agave sisalana* but is not as hardy to pests.

Agave americana

century plant
maguey
spiked aloe
bara kanwar
rakas-pattah
cabuyo
penco negro
chachuar o yana cahuar
mescal

Tropical America \ suits arid climate but will flourish in moist climates to about 1500m altitude \ tolerant of drought \ very hardy to frost down to minus 12^0C \ well armed with a spike at the point of each sharp, saw-toothed leaf \ stemless rosettes (to 3m) producing offsets \ flower spikes can grow to 8m \ leaves are greyish green although variegated varieties have yellow or white stripes \ flower buds are eaten baked, they are called *quiote* by the Kickapoo tribe and are said to taste similar to asparagus and sometimes reach diameters in excess of 50cm \ the stem base can be roasted and is sweet \ the flower head can be removed (some tribes pulped the heart) to yield a sap (aguamiel, meaning honey water) which is made into a sugar, vinegar, or is fermented into a drink called pulque, this can be distilled to mescal (mesquite), from which the Mascalero Apache tribe of California are believed to have derived their name \ an agave plant can yield between 300 to 1000 litres of *aguamiel* and then it dies \ Apache used heart and tubers to make a fermented drink \ leaf sap to treat burns and as a laxative, a diuretic, and an antiseptic \ leaf fibre soaked for 24 hours in water is used as a disinfectant for scalps \ saponin and steroids from roots \ root used to make a tea to treat arthritis \ plant also used as a treatment for ascites, leukemia, cancer, tumour, syphilis, scurvy, warts, sores and in veterinary uses for sprains and bruises on animals \ used as a depurative and cicatrizant \ leaves and roots yield a soap substitute \ a soap that lathers well in salt or fresh water is made by drying leaf juice mixed with ash, then pressing into cakes \ leaves for paper \ the extremely durable fibre *pitta thread* is made from the leaves and roots and is said to be superior to hemp \ Papago used two-ply fibre as nets for carrying \ bundles of fibres were used as hair brushes \ dried leaf was smoked by local tribes and also used as tinder \ spines on ends of leaves were used for pins or needles \ flower stalk tops used to make a ceremonial black body paint \ leaf sap is impregnated into plasters and wall papers to render them termite proof \ dried flower stems make a thatch said to be impervious to water \ leaves split and used as weft of wrapped weaving in house frames \ the size of the plant and the spines make it well suited as a living fence plant \ propagation by seed or root division \ an elegant variegated version of this species is popular in cultivation and can probably be used in the same manner.

Agave angustifolia
Agave wightii

mescal bacanora
mescal agave
dwarf sisal

Mexico and Guatemala \ small (usually 1m but can grow to 2m tall), fast growing \ frost tolerant \ grows in areas with mean annual precipitation is as low as 235mm and survives for years with annual rainfall as low as 75mm \ flowering in 8-15 years (as low as 6 years under cultivation) \ source of sap to make *mescal bacanora* which is preferred in Sonora to tequila \ the wild harvesting of more than half a million plants annually is threatening the species \ pit-roasted heart eaten \ good animal barrier \ low fibre content although the fibre is hard and long \ it is being used to breed hybrid plants which produce a superior fibre \ leaves comprise 5% steroids \ leaf extract promotes bacterial growth to reduce sewage odours in liquid manures \ pollination usually by bats.

Agave asperrima
South west USA to Mexico \ juice used as a sweetener.

Agave atrovirens
Agave salmiana
Agave quiotifera

maguey manso
metl
moguei
maguey cenizo
South western North America, Mexico \ survives arid, eroded soil \ young flower stalks cut for quiote. flower stalk is roasted and eaten as *quiote* which is chewed like sugar cane \ sap as syrup \ also used to manufacture pulque and mescal from the sap \ petals eaten in soups \ heart eaten and said to contain the rare vitamin B12 and also diosgenin \ heart is cooked and juiced to make spirit called mezcal or mescal (2 million litres in Mexico in1960) \ leaf fibre is a by-product \ leaves contain hecogenin, a steroid \ variety *salmiana* is grown exclusively for production of pulque \ outer layer of inner leaves are peeled to form a translucent food wrap \ leaves fed to dairy cattle \ popular hedge and

animal barrier \ propagation from seed or root division.

Agave bovcornuta
Young flowers eaten cooked in batter.

Agave brachystachys
Guatemala \ shampoo.

Agave brevispina

galatas
Haiti \ antiseptic \ used to treat sores, dysentery and diarrhoea.

Agave cantala

cantala
maguey
Mexico, Central America \ a very variable species \ heart of stem eaten raw or steamed \ juice is sugary for pulque \ source of fine, white, flexible fibre suited to spinning \ produces about 3 tonne of fibre per hectare at 2m x 2m spacing \ harvest of 3000 kg per hectare per year of pure fibre for about 3 years, some 4 or 5 years after planting \ fibre marketed as sisal \ harvesting is more difficult because of spiky leaves \ used for animal barrier \ propagation from seed or root division.

Agave capensis
Leaves rich in sapogenins.

Agave cerulata
Emergency water source \ wine production \ brushes made from leaf fibre \ leaves rich in sapogenins.

Agave cocui

cocuy
Venezuela \ large plant (to 3m) similar to *Agave americana* \ greyish green leaves, reddish-brown marginal spines, with edible heart \ flower buds eaten

\ alcoholic drink \ used to treat tumours.

Agave colorata
Wine from crushed stems.

Agave compluviata
Mexico \ used to make a liqueur.

Agave deserti

mescal
desert agave
century plant

South west North America to Mexico \ to 0.5m tall \ flower spike to 5m \ good drainage required \ tolerant of minimum temperatures from minus 4^{0}C to minus 12^{0}C depending on provenance \ leaf base (heart) eaten roasted \ young flower stalks eaten roasted and have the flavour of pineapple and bananas, older flower stalks very fibrous but still tasty \ Cahuilla parboiled flowers to release the bitterness, they were then eaten or dried for future use \ flour from seeds \

very young leaves were occasionally eaten raw \ leaves were baked and eaten or dried and stored \ Cocopa traded baked crowns to several other tribes \ roasted, pounded leaves and stalks were eaten and also dried and made into cakes \ mescal leaf bases were collected by Californian tribes who cooked, ate, and preserved them as a staple food \ they were cooked in earth ovens about 0.5m deep and often as wide as 3m \ the fuel was carefully selected so as not to give a bad taste to the food \ the remaining brown juicy mass was sweet and nutritious \ they were then eaten or worked into cakes and dried for winter storage \ the dried cakes were an article of trade \ the cake was eaten dried as a sweet, recooked or made into a sweet drink by boiling pieces in water \ pulque from the sap \ nectar drunk straight from the plant \ fibre is extracted from the dry leaves by beating and from the fresh leaves by soaking and rotting off the pulp and outer skin \ the strongest fibre is taken from the dead leaves \ very popular as bowstrings, rope and cord \ pounded leaf fibre was made into cactus bags, sandals, shoes, women's skirts, saddle blankets, slings,

nets, cordage and cleaning brushes \ burned plant used as the dye for tattooing, the bluish-black patterns being pricked in with the thorn of the same plant or an opuntia cactus thorn \ dried stalks for firewood \ thorns used as awls in basket making.

___

Agave difformis
Leaf juices used as soap.

___

Agave fourcroydes

henequen
Yucatan hemp
Mexican sisal

Mexico \ grows in arid to fairly humid conditions \ the plant stalk, growing to 1.8m in the wild state, averages about 0.9m under cultivation \ its grayish green, lance-shaped leaves, up to 1.8 m long and 100-150mm wide at the widest point, grow directly from the stalk, forming a dense rosette, and are edged with thorns \ flower stalk to 6m, bears greenish white flowers about 76mm across and with an unpleasant odour \ varieties of *A. fourcroydes* include *ixtli, longifolia, minima,* and *rigida* \ leaves yield 3-4% fibre, a source of textile fibre since pre-Columbian times \ first shipped from the port of Sisal in 1839 \ introduced to Cuba in the 19th century, becoming the country's chief fibre crop by the 1920s \ probably the plant most

used to produce sisal fibre in Mexico where it is preferred to *Agave sisalana* \ fibre harvesting commences when the plant is 5-7 years old and continues for about 15 years \ planted at about 3,000 plants per hectare \ in Mexico in 1970 there were 200,000 hectares planted \ the fibre yield in Mexico in 1975 was 139,000 tonnes \ fibre is sometimes referred to as Yucatan sisal or Cuban sisal \ plants yield about 25 leaves annually from about the 5th through to the 16th year after planting \ outer leaves are cut off close to the stalk as they approach their full length \ commercially, the fibre is freed by machine decortication, a process which crushes the leaf between rollers and scrapes the resulting pulp from the fibre \ fibre strands are then washed, dried in the sun, and brushed \ fibre strands are lustrous, white or yellow and average about 1.2 to 1.5 m in length \ fibre has reasonable resistance to deterioration from micro-organisms found in salt water and fairly good strength and ability to stretch \ henequen fibre is made into twines used in agriculture and shipping and is also made into rope, bags, hammocks, and shoe soles \ possible use of waste fibre for paper pulp \ leaf waste yields about 2% wax \ wastes also can be made into industrial quality alcohol \ considered inferior to *Agave sisalana* because the leaf edges are very prickly and hard to handle \ detergent from leaves \ hecogenin from leaf juices while extracting fibre \ leaves fed to dairy cattle.

Agave funkiana
Jaumave fibre
Jaumave valley of Mexico \ high altitude \ produces a fibre which withstands hard wear and is water resistant \ often used in brushes.

Agave ghiesbreghtii
Popular hedge and animal barrier.

Agave goldmaniana
South west North America \ used as food - parts unknown.

Agave gracilispina
Mexico \ used to make a liqueur.

Agave hookeri
Used as hedge and animal barrier.

Agave intermixta
galata
maguey
Dominican Republic \ used as a laxative and a diuretic.

Agave kirchneriana
Mexico \ used to make a liqueur.

Agave latissima
Mexico \ used to make a liqueur.

Agave lechuguilla
Istle
tula
lechuguilla
Tampico hemp
Mexican fibre

Mexico, New Mexico and Texas \ small, open rosettes and straight leaves \ very drought tolerant but will flourish in moist climates to about 1500m altitude \ very hardy with some provenances often snow covered for weeks and able to withstand temperatures as low as minus 23⁰C \ grows in limestone mesas and along hillsides \ plants mature in 8-10 years then flower and die \ young shoots boiled and eaten \ produces a fibre from 200-450mm long which is stiff and can be bleached, withstands hard wear and is water resistant \ often used in brushes (including hair brushes), but also in sacking, upholstery tow and cordage \ broken fibre used in a fibre board \ leaves and stems rich in saponins (about 1%) and used as soap

substitute and fish poison \ leaf extract used to poison arrow tips \ steroids in leaves and roots including smilagenin \ 3.7% hecogenin in green fruit \ seed contains manogenin and hecogenin \ used as a snake bite treatment.

Agave letonae
Salvador henequen
letona fibre
El Salvador \ fibre is used.

Agave mapisaga
Mexico \ large plant with wide, thick, fleshy leaves \ abundance of *aguamiel* is used to make a pulque \ outer layer of inner leaves are peeled to form a translucent food wrap \ popular hedge and animal barrier.

Agave melliflua
Mexico \ used to make a liqueur.

Agave mescal
Cultivated in Jalisco in Mexico \ used to make tequila and other alcoholic drinks \ leaves as by-products are possible source of fibre and sapogenins.

Agave mexicana
South west North America to Mexico \ baked roots are eaten.

Agave nelsonii
Leaves rich source of sapogenins.

Agave neomexicana
South west North America \ mountain areas \ very hardy with some provenances often snow covered for weeks and able to withstand temperatures as low as minus 23⁰C \ drought tolerant \ roasted pulp can be stored for when needed.

Agave palmeri

mescal
Palmers agave
century plant

South west North America \ slower growing than many Agave species \ very hardy to frost down to minus 12^0C \ central buds eaten roasted \ sweet, juicy food from leaf base \ heart of crown eaten by children as a candy \ plant eaten dried \ used as an emergency food \ flower stalks baked and chewed for juice \ pulque from sap \ Apache fermented the cooked crowns in a vessel, ground and boiled the contents, then fermented the liquor again \ heart juice was also strained and mixed with "*tiswin* water" a liquor made from fermented maize \ fibre from leaves used locally to make ropes \ thorn used as a needle for thread \sap of the agave plant has disinfectant properties and can be taken to check the growth of putrefactive bacteria in the stomach and intestines \ that water soaked in agave fibres is claimed to be useful in the treatment of falling hair young \ Apache girls daub juice on their cheeks \ the juice left on the pit stones after baking was used to paint stripes on buckskin \ stalks were made into lances and hoe handles.

Agave parryi

century plant

Arizona and New Mexico \ to 0.5m tall and 1m spread \ stiff, broad, grey-green leaves with a single spine at the pointed tip \ flower stem to 4m tall bears creamy yellow flowers \ very hardy to cold and drought \ with some provenances often snow covered for weeks and able to withstand temperatures as low as minus 23^0C \ best in well-drained, sunny position \ once a staple food of the Apaches \ bulbous crowns were baked in pits and the pulpy centres released and pounded into thin sheets for eating \ pit cooked leaf bases often made into cakes, dried and stored as food \ young flower stalks and young leaves are eaten roasted \ stalks eaten raw, boiled or roasted \ boiled stalks often dried as stored vegetable \ flower stalks baked and chewed for the juice \ heart of the crown eaten by children as a candy \ syrup made from nectar \ roots eaten baked \ Apache fermented juice as a drink \ juice strained and mixed with *tiswin* water \ Apache cooked the crowns which they fermented in a vessel then ground and boiled the resultant mix and allowed it to ferment a second time \ thorn used as a needle for thread \ young Apache girls daubed juice on their cheeks \ juice covering pit stones was used to paint stripes on buckskin \ stalk used for lance shafts and hoe handles \ item of trade between tribes \ propagation from seed.

Agave patonii

South west North America \ staple of some tribes \ roasted flower stalks are edible.

Agave pes-mulae

Mexico \ used to make a liqueur.

Agave potatorum

Mexico \ in Sonora the hearts are placed in subterranean ovens and the resulting fermented juice is distilled to make a spirit called bacanora.

Agave promontorii

Mexico, south west USA \ source of fibre and rich in sapogenins.

Agave pseudotequilana

Mexico \ used to make a liqueur.

Agave quiotifera

Used to make a liqueur.

Agave rhodacantha

Mexico \ commercial leaf fibre from leaves.

Agave rigida

henequin

Good fibre.

Agave rubescens

Mexico \ fruits and flowers eaten cooked.

Agave scabra

South west USA and Mexico \ hardy to frequent, moderate frost to minus 8^0C \ tender young stems are eaten \ leaf bases are eaten \ central bud is eaten roasted \ a spirit is made from the heart.

Agave schottii

amole
soso
Schott agave

South west Arizona, New Mexico and Mexico \ very hardy to frost down to minus 12^0C \ roasted stem pith is eaten \ crushed leaves yield a soap substitute and fish poison \ potential source of sapogenins \ flower spikes as arrow shafts.

Agave shawii

mescal
Shaw's agave

South west North America \ roasted stem pith is eaten \ uses similar to *Agave deserti*.

Agave sisalana

sisal hemp
bans kawara
perrine
yaxci

Mexico but now Pantropical \ this species is probably a result of selective breeding from other species over a long time frame \ few prickles on leaves \ plant flowers after 7 to 10 years and throws a flower spike to 7m tall \ these flower spikes do not always produce seed, but about 2000 bulbils are produced on a stem \ bulbils are preferred to suckers because suckers are less vigorous, flower at the same time as the parent plant and do not provide uniform growth \ best grown on well-drained limestone soil in a hot, dry or moderately wet climate (650-1300mm rainfall preferred in India) \ planting density about 1500 plants per hectare \ pulque and mescal are made from the sap of the flowering stalk \ heart eaten roasted \ used as a cicatrizant, depurative, sudorific and detergent \ used to treat syphilis, leprosy and dysentery \ thick succulent leaves from 1.3-2m long has provided about half the world's natural hard fibres \

fibre drying after processing
leaf fibre is source of sisal fibre which is a valued alternative to Manila fibre and to jute \ it has a high breaking strain and low extensibility \ used to make heavy twines, ropes and marine cordage, mats, sacks, tarpaulins, mops, brushes, kraft paper, cardboard, and the waste fibre is used to make cheap twines and upholstery tow \ first leaf cut when the plant is about 1.5m tall and bears about 100 leaves, usually when the plant is 2 to 4 years, depending on soil types (and other growth regulating factors) \ the bottom 35-40 leaves are cut

yielding about 900kg per hectare of fibre \ annual cuts after this yield from 2000-2800 kg fibre per hectare until the plant goes to flower \ typical fibre content of leaves is 3% in wet climates and 4% in dry climates \ wastes from fibre production contain wax and hecogenin for steroid production (0.6-1.3% depending on age of the plant) \ hedge and barrier plant \ hybrids have been developed in East Africa which have superior fibre but lower hecogenin content \ soap.

Agave sobolifera
cocui
Venezuela \ used to treat tumours.

Agave sobria var. roseana
Agave roseana
Mexico \ rich source of sapogenins (especially hecogenin in leaves - highest of all *Agave spp.* 2.5%).

Agave stricta
hedgehog agave
Mexico \ young flowers eaten cooked in batter \ commercial leaf fibre from leaves.

Agave sub-simplex
Sonoran desert \ flower spikes used to pick cactus fruit, necklaces from flat black seeds \ face paint from stems.

Agave subtilis
Mexico \ used to make a liqueur.

Agave sullivani
Mexico \ used to make a liqueur.

Agave tecta
Used as a hedge plant and to protect crop from animals.

Agave tequilana
Agave azul var. tequilana
blue agave
chino azul
mescal
tequila agave
zapupe

Mexico \ plants grown at one metre spacing in rows 3m apart \ heart is rich (14.3-24.1%) in inulin (it is likely that other species are also rich in this valuable low calorie sugar) \ hearts are harvested, split in half, cooked by steaming and used to make tequila \ clones of *Agave tequilana Weber azul* are usually used for tequila production \ a distilled liquor, usually clear in colour, unaged and contains 40-50 percent alcohol is made from the sweet sap, or *aguamiel* (honey water) \ by law at least 51% of raw materials to make tequila must be from this species, the remainder being made from other agaves or sugar cane \ juice is fermented and then distilled twice to achieve the desired purity \ some brands are aged in oak vats, which allows the distillate to mellow and take on a pale straw colour \ fibre from leaves is a by-product and is softer and finer than that of *Agave fourcroydes* \ wounds from the thorn are treated with juice from the leaf of the same plant.

Agave toumeyana
Arizona \ very hardy to frost down to minus 12^{0}C \ leaf juice used as soap.

Agave utahensis
Utah aloe
mescal
Western North America \ erect, compact grey-green rosette with tapering, stiff leaves with spines up the margins and terminating with a long spine at the tip \ to 0.25m tall and wide \ usually forms clumps about 2m across \ flower stem to 1.5m long \ very hardy, often snow covered for weeks and able to withstand temperatures as low as minus 23^{0}C \ uses similar to *Agave deserti* \ flowers are boiled, dried and stored as food \ bulb of the root is roasted as a delicacy and said to be sweet and delicious \ short stalks, buds, short flower stalks and leaf bases are eaten baked and a beverage is made from soaking this in water.

Agave vera-cruz
Agave lurida
maguey
East Indian agave
Mexico, naturalised in India and other parts of Asia \ large plant with broad, thick, spiny leaves often very curved \ wax gives them a grey-green colour \ in Mexico pulque production at year 8-10 \ pilot plant in India is extracting 10% fructose from the hearts (this implies the heart is rich in

inulin), ie 8.75 tonnes of fructose per hectare (probably from a 5 year old plantation) \ edible starch from plant heart which is also used to make tequila \ as a by-product of tequila making, fibre obtained from the leaf of the plant (1.5-2.5%) \ fibre is shorter and stiffer than that of *Agave fourcroydes* \ although the fibre is strong and white it is considered to be inferior with physical properties similar to the hard leaf fibre cantala \ used for rope and cordage \ a simple hand spindle known as a malacate is used to spin fairly fine yarn from the maguey and related hard fibres \ in South America the name maguey is used for a variety of fibres as well as for the plants from which they are derived.

Agave Victoria-reginae
pintillo

North east Mexico where it is endangered in the wild (CITES listed) \ small, low-growing plant which does not produce suckers \ very hardy to frost down to minus 12^{0}C \ requires alkaline soil \ popular as ornamental and house plant \ fibre called "noa".

Agave vilmoriniana
octopus agave
midas agave

Sonora, Mexico \ blue-grey, twisted leaves to 1.2m long and 200mm wide \ tolerant of shade \ space about 2m apart for mass planting \ leaf juice traditionally used as a soap substitute \ mature leaves contain up to 4.5% sapogenins (7-8 year old plants).

Agave virginica
Used to treat spasm and colic \ used as a stomachic.

Agave vivipera
Mexico \ young stems as emergency food \ pulque production at years 8-10.

Agave weberi
smooth-edged agave

Mexico \ leaves to 1.2m long and 200mm wide without saw-tooth edges, just a spike at the tip \ very cold tolerant \ sensitive to light, closing up when it is too bright or too hot \ used to make a liqueur \ used as a hedge plant and as an animal barrier \ propagation

from seed or offsets at the base.

Agave wislizeni
Mexico \ tender young shoots are relished.

Agave wocamhi
Mexico \ edible flowers.

Agave yaquina
su'ut
maguey
Mexico \ in Sonora the hearts are placed in subterranean ovens and the resulting fermented juice is distilled to make a spirit called bacanora \ leaf fibre for baskets.

Agave zapupe
Mexico \ plants used for pulque at about age 8-10 \ leaf fibre used as henequin fibre in Mexico \ leaves contain hecogenin, tigogenin, and gentrogenin.

Cordyline australis
ti-kouka
whanake
giant dracaena
cabbage tree palm

New Zealand \ evergreen tree to 12m and spreading to 3m \ part shade or full sun \ frost hardy \ young leaves and shoots are eaten raw or roasted \ large fleshy rhizome can be eaten boiled or brewed into an intoxicating drink \ leaf fibre is strong \ propagation from seed and cutting.

Cordyline fruticosa
Cordyline terminalis

**ti palm
la'i
andong
Hawaiian red ti
palm lily
good luck plant
tie shu ye
andong
daoen nagasi
daun juang-juang
jejuang
jenjuang
laklak
lenjuang
senjuang
tunjung
masawe
que
vasili
ta'un
ariko
tubui
kokotodamu
si
rau ti
'auti ti
ki
la'i
ting
kava
kaut bu
bauga
elaivi**

North eastern Australia, Pacific Islands, tropical Asia \ long leaved plant on a straight trunk (to 5m and spreading to 1.5m) \ not tolerant of frost \ tender leaves eaten as a flavouring to rice \ wrapped around fish prior to baking \ roots which can grow in excess of 100kgs are rich in sugar and eaten cooked \ cooked roots can be fermented into a drink called *okolehao* \ rhizome treats diarrhoea \ traditional Polynesian skirts (*ti ti*) made from the leaves \ leaves also used for thatch and to make sandals \ juice used to treat sore or infected eyes \ infusion taken in early part of pregnancy very effective for abortion \ leaves used as a contraceptive \ macerated roots to cure toothache and laryngitis \ outer rind of flower stalk valued for anti-syphilitic properties \ root decoction given to mothers when their milk turns yellow and causes child to vomit and have diarrhoea \ young leaf buds used for chest pains \ new shoots taken as a remedy for filariasis \ liquid from stem taken to relieve sickness after childbirth and to aid expulsion of afterbirth \ roots for baldness \ mature leaves crushed with oil and applied abscesses of the gums \ used to treat abscesses, ear ache, amenorrhea, asthma, gastritis, eczema, whooping cough, debility, fever, inflammation, polyps, pyorrhoea, smallpox and wounds \ used as a laxative, aphrodisiac, sweetener and fish poison \ used to treat stomach ailments, hepatoma, lung tumour, dermatosis, ague (malarial fever) \ cough, insanity, dyspepsia, smallpox, arthritis, diarrhoea, gingivitis and wounds \ propagation from seed.

Cordyline manners-suttoniae
palm lily
Queensland \ roots eaten after leaching and boiling \ juice from roots used as an antiseptic and contraceptive.

Dasylirion leiophyllum

Mexico \ used to treat toothache.

Dasylirion longissimum
Mexico \ used to make distilled alcohol.

Dasylirion texanum
Texas sotol
Texas \ hardy to frequent, moderate frost to minus 8^0C \ leaves and young stems are eaten cooked \ central part of the bud is eaten roasted and also made into a beverage \ crowns pit baked, dried pounded into flour and made into cakes.

Dasylirion wheeleri

Wheeler sotol
desert spoon
arbita maagama
tacut
palma
wechesas
yerey palma
palma del suelo
zamij etea

Sonoran Desert, Texas, Mexico \ grows to 2m tall \ long, graceful, grey-green, ribbon-like leaves grow to 1m long and are edged with hooked spikes \ older plants develop a short trunk and several heads \ flower spikes to 2m long \ plant does not die after flowering \ very hardy to frost down to minus 12⁰C \ plant in full sun and give minimal water \ central bud eaten roasted \ pit baked crowns stripped, pounded to a pulp, spread out and dried, then eaten like cake \ fresh young stalks eaten raw, roasted or boiled \ boiled stalks can be dried and stored as a vegetable or pounded as a drink \ head hearts cooked with bones to make soups \ crowns pit baked, removed, peeled, crushed, mixed with water, fermented and used as a beverage \ a source of alcohol \ seed is 26%

protein, 22% oil \ leaf fibre as cordage \ stalks used as cross pieces for cradle-board backs \ leaves twilled into mats and used in baskets and clothing \ leaf bases and stalks used to make cigarette papers \ leaves pulled away at the base can be used as a spoon.

Dracaena spp.

Genus of evergreen shrubs and trees, popular as ornamentals \ prefer full sun and well drained soil \ plants can be cut back without killing \ propagation from seed or cutting.

Dracaena arborea

niõmmé
abé

Western Africa \ planted as boundary fences.

Dracaena angustifolia
Pleomele angustifolia

semar
sudji

Tropical Asia \ young leaves eaten cooked with rice \ pounded leaves mixed with water as a green food colouring \ fruit eaten roasted.

Dracaena aurea

Hawaii \ used to treat chill, fever, asthma and other lung complaints.

Dracaena cinnabari

socotra dragon's blood

Egypt \ used as an astringent and as a cicatrizat \ used to treat diarrhoea, dysentery and haemorrhage.

Dracaena congesta

jenjuang
juang-juang
lenjuang
senjuang

Malaysia \ used to treat rheumatism and sores \ used as a vermifuge and anodyne.

Dracaena cylindrica

Western Africa \ boundary and fence plant \ leaves for fibre.

Dracaena deremensis

Janet Craig
warneckei

Tropical Africa \ very tolerant of low light conditions \ one of the known indoor air filtering plants and the variety 'Janet Craig' has been found to be especially good.

Dracaena draco

dragon tree

Canary Islands – now very rare in the wild \ evergreen tree to 20m tall \ stem may grow as thick as 4m diameter \ best in a frost-free environment \ yellow-orange fruit is eaten ripe \ source of dragon's blood a resin which oozes from the trunk which is used in cooking, varnishes, dyes, paints, lipsticks \ propagation from seed or cuttings.

Dracaena fragrans
Pleomele fragrans
Dracaena smithii

cocked hat
happy plant
corn plant

Ghana to Cameroon \ closed forest \ small, solitary, erect tree to 6m with a 2m spread \ good in coastal regions \ fibre \ boundary fences \ leaves are sword-shaped and glossy and are harvested for fibre \ few drops of sap applied to breast act as baby purge \ juice mixed with palm oil as a rub against fever \ roots against stomach pain \ bark decoction as lotion for rheumatism \ one of the known indoor air filtering plants and is especially useful in the removal of formaldehyde \ propagation is easy from cuttings.

Dracaena graminfolia
ramput
julong
Used as an antidote.

Dracaena mannii
vulture's screw pine
okwasu-ampa
kesene
kesenekesene
kesrekesre
soap tree
Senegal to Cameroon \ tree to 12m \ core of immature leaves eaten as asparagus (common food of gorilla) \ tree commonly planted as a hedge \ soap made from leaves \ fibre from retted leaves \ yellow dye from the stem.

Dracaena marginata
dragon tree

Madagascar \ tolerant of dry and low light conditions \ one of the known indoor air filtering plants and is especially good for removing xylene trichoethylene.

Dracaena ovata
budio
Sierra Leone \ common understorey plant of forest \ grown as a boundary plant and fence.

Dracaena perrotteti
Ghana \ used to treat syncope.

Dracaena porteri
jenjuang
lenjuanjuang
senjuang
juang-juang
Used as a an anodyne and a vermifuge \ used to treat rheumatism and sores.

Dracaena scoparia
Ivory Coast to Ghana \ shrub or small tree with many branches \ plant is difficult to kill and used a lot in living fences.

Dracaena surculosa
kwae-beten
mobia
gold dust dracaena
Sierra Leone to Ghana \ suckering, multi-branched climber or small tree \ popular greenhouse ornamental \ yields a red dye \ swollen roots cut up and boiled with tiger-nuts *(Cyperus esculentus)* as an aphrodisiac \ propagation from seed or stem cutting.

Dracaena umbratica
Used to treat rheumatism.

Furcraea spp.
cabuya
pita
Genus of perennial succulents similar in appearance to agave \ best in full sun and well-drained soil \ fibre for cordage \ fish poison \ pounding leaves and flower stems yield a detergent with bleaching qualities.

Furcraea agavephylla
Trinidad \ depurative \ used to treat sores and rheumatism.

Furcraea andina
Peru, Ecuador \ cultivated locally for fibre

Furcraea cabuya
Costa Rica \ cultivated locally for fibre

Furcraea foetida
Agave gigantea
Furcraea gigantea

cocuiza
French aloe
green aloe
Mauritius hemp
Portuguese pitiera

Venezuela, Brazil \ introduced to Mauritius in the late 18th century \ large (2-3m) succulent perennial with large fleshy leaves \ at maturity the flower stem can grow to 6m \ it resembles agave in appearance \ it has become naturalised in many tropical areas \ the grey-green, lance-shaped leaves (1.2 to 2.1 m long and about 8 inches (20 cm) across the widest portion and sometimes edged with thorns) grow directly from the short plant stalk to form a dense rosette \ flower stalk, which appears near the end of the plant's life span, about 8 to 10 years after planting, grows up to 12.2 m and bears white flowers about 3.8 cm long \ produces leaves suitable for harvest within 3 to 4 years after planting and each 18 to 36 months thereafter, yielding about 25-30 leaves at each harvest \ leaves undergo decortication a scraping process that is sometimes preceded by several days of retting \ processing is completed by washing and drying \ fibre is sometimes brushed, which adds softness and lustre \ fibre not as strong as the leaf fibres sisal and henequen but is softer and finer \ careful processing of the fibre strands, about 4 to 7 feet long, yields creamy-white fibre with fair lustre \ it is easily died and is fairly resistant to deterioration in fresh water but is subject to damage in salt water \ made into bagging and other coarse fabrics and is sometimes mixed with other fibres to improve colour in rope.\ fibre from the highland is referred to as aloe Malgache and from the lower areas as aloe Creole \ cultivation was established in East Africa, Ceylon (now Sri Lanka), and St. Helena late in the 19th century \ fibre has been grown commercially in Mauritius and is used for ropes, twine, sacks and mats \ planting is less than 2m x 2m in light soil and more than 2m x 2m in heavy soils \ used to treat tumours \ in India it is used as a fish poison \ propagation is from bulbils produced on the flower stem or from root suckers.

Furcraea hexapetala
bayonette
bois pitre
henequen
jenequen
pite franc
pite pays
pitre pays

Jamaica, Trinidad, Dominican Republic \ cultivated locally for fibre \ used as a treatment for lip cancer \ to treat hypochondria \ cataplasm \ astringent \ diuretic \ to treat dysentery \ wounds in animals \ soap \ fish poison.

Furcraea humboldtiana
cocuiza

Venezuela \ cultivated locally for fibre \ used as a depurative and to treat tumours.

Furcraea macrophylla
Colombia \ cultivated locally for fibre

Furcraea tuberosa
cabuya
cocuiza
maguey
pite espagnole
pite

Dominican Republic \ diuretic \ used to treat wounds in animals \ fish poison \ soap.

Nolina matapensis
leaf fibre for weaving baskets.

Nolina microcarpa
palma

used for manufacturing cheese-making baskets.

Phormium cookianum
Phormium colensoi

New Zealand flax
mountain flax

New Zealand \ evergreen perennial to 2m tall with sword-shaped, upright, dark green leaves spreading to 1m \ yellowish-green flowers \ requires sunny site with moist, well-drained soil \ hardy to frost and drought \ plant used to treat tumours and tropical ulcers \ leaf fibre used \ propagation from seed or division.

Phormium tenax
New Zealand flax
harakeke

New Zealand \ evergreen, upright perennial to 3m tall \ orange and red flowers \ coastal to sub-alpine, mostly in lowland swamp \ frost hardy \ best in sunny situation in moist, well-drained soil \ good coastal plant \ nectar is eaten and relished \ in New Zealand the roots were boiled and used as an abortifacient \ plant used to treat tumours, cancer, tropical ulcers and used as a purgative and vermifuge \ leaf fibre used for clothing, matting, baskets and sandles.

Polianthus tuberosa
tuberose

Flowers eaten in soups \ used in manufacture of kekap, an Indonesian soy sauce.

Sansevieria spp.

Genus of rhizomatous, evergreen perennials \ produce few leaves clumped on long root systems which throw out new leaf clumps along their length \ generally require fairly dry winters with minimum temperatures no lower than 10^0C \ indirect sunlight \ thrive in warm, humid conditions \ they are popular as house plants where they prove very hardy \ many species harvested locally for leaf fibres which are used to make bowstrings, rope, cordage, hats, matting and clothing \ medicinal uses for sap, rhizomes and powdered or ground roots \ propagation from seed, leaf cuttings or division in warmer weather.

Sansevieria angolensis

Western tropical Africa \ leaves contain fibre used for making deep sea dredging ropes.

Sansevieria cylindrica
ife hemp

Angola \ stemless with cylindrical, erect leaves 1-2m long \ plant is slow growing \ yields a fine, white fibre \ leaves contain 1.5% fine white fibre \ yields to 670kg/hectare/year \ propagation from sucker or seed.

Sansevieria enrenbergii

Somalia \ arid climate \ slow growing \ leaves yield 1.5% of fine white fibre \ yields up to 670kg of fibre per hectare per year.

Sansevieria gracilis

Southern Africa \ edible flowers.

Sansevieria grandis
grand Somali hemp

Tropical Africa, Somaliland \ running rhizomes produce up to 5 broad, obovate leaves to 1.2m long and 150mm wide \ good fibre.

Sansevieria hyacinthoides
Sansevieria guineensis
konje hemp
African bowstring hemp
mapana
snake plant

Southern Africa \ stemless, stout creeping rhizome producing 8-10 flat, sword shaped leaves from 1-2m long and 100mm wide \ leaves produce a good fibre at about 2% extraction rate \ yields up to 2250kg fibre/hectare of land \ used to treat fracture \ cultivated in Las Perlas in Central America where it is regarded as a snake bite cure \ best propagated from leaf tip cuttings.

Sansevieria intermedia
pygmy bowstring

East tropical Africa \ 2-7 leaves from 0.45-1.2m long and 20mm thick \ on a stemless plant \ fibre from leaves.

Sansevieria liberica
chanvre d'Afrique

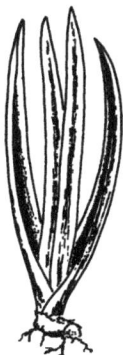

Tropical west Africa \ stemless, creeping rhizome with 1-3 or more leaves in a clump \ 1500kg of fibre per hectare per year \ fibre used for fishing lines, nets, bowstrings, shoes and as a sponge for infants \ it is also woven into cloth \ raw leaf pulp applied to ulcers and smallpox sores \ raw juice applied directly for eye and ear problems \ fluid an ingredient in toothache remedy \ leaves burned and smoke inhaled for feverish headaches \ root as a stimulant and tonic.

Sansevieria longiflora
1500kg of fibre per hectare per year \ fibre for mats, twines and bowstrings.

Sansevieria roxburghiana
manjinaru
saga
Indian bowstring hemp

India \ short stem or stemless with up to 9 leaves to 0.75m long and 25mm wide in a clump on a creeping rootstock which may bear up to 30 clumps \ used as a snake bite treatment, haemolytic, vermifuge and parasiticide \ leaves are source of commercial *murva* or *moorva* fibre which is soft, silky, pliant and very elastic \ especially prized in bowstrings \ also used for ropes (used for deep sea dredging), thread, twines, matting \ woven fibre makes fine clothes which can be dyed \ leaves good for paper pulp.

Sansevieria scabrifolia
Africa \ used by bushmen as a water source.

Sansevieria senegambica
African bowstring hemp
Tropical West Africa \ stemless creeping rhizome \ 3-4 leaves (0.7m long and 60 mm across) per clump \ fibre used for fishing lines, nets, bowstrings, shoes and as a sponge for

infants \ it is also woven into cloth \ raw leaf pulp applied to ulcers and smallpox sores \ raw juice applied directly for eye and ear problems \ fluid an ingredient in toothache remedy \ leaves burned and smoke inhaled for feverish headaches \ root as a stimulant and tonic.

Sansevieria trifasciata
bowstring hemp
zoreil
bourik
snake plant
leopard lily
mother in law's tongue

Western tropical Africa \ stemless, erect, stiff, grouped or solitary plant with leaves to 1.2m long and 65mm wide are used for fibre \ fibre used for fishing lines, nets, bowstrings, shoes and as a sponge for infants \ it is also woven into cloth \ raw leaf pulp applied to ulcers and smallpox sores \ raw juice applied directly for eye and ear problems \ fluid an ingredient in toothache remedy \ leaves burned and smoke inhaled for feverish headaches \ root as a stimulant and tonic \ roots for baldness, \ plant used to treat snake bite, malaria, fever, sores, earache and itch \ one of the known indoor air filtering plants \ popular as an ornamental.

Sansevieria zeylanica

nijanda
Devils tongue
Ceylon bowstring

India and Sri Lanka \ stemless rosette of leaves up to 2m long and 20mm wide \ naturally occurs dry or rocky soil at low elevations \ thrives in most moist climates up to 650m altitude \ planted in rows 600mm x 300mm commercially \ leaves, harvested after 2-3 years, contain 2% fibre \ when fully bearing, fibre production is typically 1500kg/hectare per year \ fibre is silky white and woven into bowstrings and mats and used generally in weaving \ used to treat abdominal tumours \ propagation from seed, suckers or leaf cuttings.

Yucca spp.

Hesperoyucca spp.

yucca

Genus of evergreen shrubs and trees \ the genus is variable and unstable with natural hybridization and intermediate forms occurring in the wild \ the same common name is often applied to many species \ desert survivors from south west USA and northern Mexico \ commonest in sand and gravel soils \ able to withstand alkaline soils, low rainfall and extremes in temperature \ need full sun and well drained soil \ generally slow growing. Extremely valuable to the Amerindians. Roots are source of saponin \ leaves a source of fibre \ important food plants \ construction of dwellings and household items \ plant juice used as base in liquid fertilisers \ for commercial fibre production they are slow (3-5 years) to recover sufficiently from a harvest of leaves to be harvestable again \ being rich in saponins they are of interest commercially as a source of sapogenins for manufacture of steroids for birth control and cortisone \ propagation from seed \ optimum germination between 17^0C at night and 20^0C daytime maximum, at 13-14 hours daylight per day.

Yucca acaulis

maguey

Venezuela \ young leaves are eaten \ a spirit is made from the ferment of the sweet juice of the plant.

Yucca aloifolia

Spanish bayonet
Spanish dagger
dwarf yucca

Florida, Mexico, Central America and Caribbean \ slow-growing, erect, evergreen shrub or small tree, sometimes branching and growing from 3 to 7m tall \ 500-750mm long leaves are shorter than most yuccas (300mm) sword-shaped with sharp points and form rosettes \ cold and drought hardy \ tolerates coastal exposure \ well drained soil in full sun \ flower petals are crisp and eaten raw in salads, or fried in batter \ flower stalk (20cm wide and 1m long) is eaten peeled and boiled \ buds eaten raw, roasted or boiled \ Choctaws made a salve from the root decoction by boiling root and mashing with tallow or grease \ plant used as a diuretic, purgative, soap and to treat tumours and lung disorders \ propagation from seed and root division.

Yucca arizonica

Arizona \ seeds have very high levels of sapogenins (12% sarsasapogenin dry weight) which can be converted chemically to cortisone.

Yucca australis

Mexico \ flowers eaten in salads \ used as a cathartic and to make liqueur \ trunks of bigger plants for building.

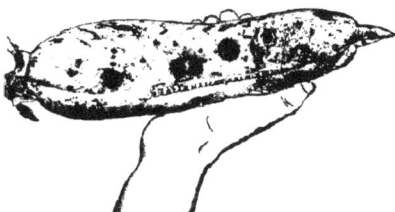

Fruit of Yucca baccata

Yucca baccata

samo'a
hoskawn
amole
palmilla ancha
banana yucca
dátil
pita
soapweed
broad leaf yucca
blue yucca
standing awl
Spanish bayonet
wild leafed yucca

South western USA and Mexico \ evergreen, clumping, woody plant \ large, upright clusters of fruit \ long lived \ found growing in canyons and on mountain slopes in full sun \ very hardy, often snow covered for weeks and able to withstand temperatures as low as minus 23^0C \ fully ripe, banana-like fruit (*p'ape*) has edible pulp which is sweet and highly esteemed with a flavour similar to dates \ sugar-rich fruits eaten raw, boiled, roasted or dried and rolled into loaves and stored for winter use \ dried fruit dissolved in water as a drink \ juice of baked fruit drunk by Apaches \ soaked, cooked fruit made into a syrup and used like hot chocolate \ seed recovered from pulp of roasted fruit and ground and formed into large cakes by the Apaches \ tender crowns roasted and eaten \ flower buds are roasted just before they expand \ flowers are boiled for food \ young leaves are baked and eaten as an emergency food \ leaves boiled with meat or cooked in soups \ a major fibre plant, the leaves being soaked then pounded on a flat rock with a wooden beater and washed until reduced to leaf fibre \ some tribes baked the leaves in earth ovens to render them sweet so that they could be chewed back to fibre \ whichever method is employed, the remaining fibre is very strong and comparable to Manila hemp \ for sandals (plaited whole leaves), ropes, twines, nets, hats, shoes, hair brushes, paint brushes, mattresses, blankets and baskets \ plants used in heartburn and to treat snake bite \ fruit eaten raw as a cathartic or to promote easy childbirth \ large roots are rich in saponins and used as a soap or shampoo with excellent lather \ roots are bruised with a stone then soaked in water for a few minutes then stirred to lather \ yucca juice used by Ramah Navaho as a hand lubricant for midwife delivering baby \ flowers have potential in perfumes \ seeds contain 6.8% sarsasapogenin.

Yucca brevifolia

Yucca arborescens
Yucca draconis var. arborescens

**Joshua tree
tree yucca
cactus yucca
yucca palm**

Mojave desert, California, Utah, Arizona, Nevada, Mexico \ major threats in California include vandalism, fire, overgrazing, and being removed for planting in residential areas (they rarely survive) \ erect, evergreen, slow-growing, small, branching tree to 9m \ leaves to 250mm long \ long lived (1000 years or more) \ grows in: extensive open pinion-juniper woodlands; desert shrub; and desert grassland \ 600-2100m altitude \ coarse sand, fine silt and bimodal soils \ natural rainfall can be as low as 130mm and occurs mostly during winter or early spring with little in summer \ very hardy to frost down to minus 12^{0}C \ drought hardy and able to withstand very high temperatures \ older plants generally survive fires and sucker extensively from their rhizomes \ seedlings will often colonise after a fire \ pollination in the wild by yucca moth, so may need hand pollinating in cultivation \ on extremely harsh sites flowering rarely occurs \ best in well-drained soils in an open, sunny position \ fleshy fruits are highly palatable \ sweet flower buds roasted on coals and eaten or parboiled in salt water and cooked again and served like cauliflower \ open buds are rich in sugar and eaten (roasted) as a candy \ seeds were ground and eaten either cooked in a mush or raw \ roots eaten raw, roasted or boiled \ seed contains 34% edible, semi-drying oil which is clear tasteless and nearly odorless \ native Americans fermented the buds and flowers to make an alcoholic drink \ a plant extract is marketed as brevifoline and is used to foam ginger beer and root beer \ plant used as an emetic and a treatment for gonorrhea \ an important wildlife plant \ cattle and sheep eat the leaves which are nutritional but have a soapy taste \ rootlets used by native Americans to make a red dye \ leaf fibre used for rope, mats, sandals and baskets \ plant was used for pulp for fine paper and for some time a London newspaper used this paper \ living trunks were incorporated into living dwellings by ancient cliff-dwellers \ wood is light, pliable, porous and durable with an attractive grain and is used for small items such as picture frame and as a veneer which improves acoustics and insulation \ chemicals extracted from the plant have been used in the synthesis of vanillin, as a fertiliser, and as a carbon dioxide stabiliser in controlling oil fires \ it is an important shade plant \ propagation from seed which can begin in just a few days \ vegetative reproduction also occurs from underground. rhizomes and root sprouts.

Yucca carnerosana

Samuela carnerosana

**palma samandoca
palm barretta**

Texas, Mexico \ long-lived (to 75 years) tree to 9m tall with an unbranched thick trunk and a crown of long narrow leaves \ from mountain crests to valley bottoms \ calcareous soils \ altitude to 3000m \ very hardy to frost down to minus 12^{0}C \

immature flower buds eaten roasted or boiled \ fibre from inner leaves (known as *ixtle de palma*) can be bleached, withstands hard wear and is water resistant \ often used in brushes, but also in sacking, upholstery tow and cordage \ broken fibre used in a fibre board \ pulp can be made into kraft (brown) paper \ residues can be used for soap \ trunks used in house construction and fences.

Yucca constricta
Texas \ seed contains 26% semi-drying oil.

Yucca elata

palmella
beargrass
soapweed
soap tree yucca

South western USA, Mexico \ creamy-white, lily-like flower is the floral emblem of New Mexico \ tree-like yucca sometimes growing to 6m tall \ blue-grey, grass-like leaves to 800mm long and only 10mm wide \ flower stalks are the tallest of the yuccas, growing to 2m long in late spring / high desert plant \ very hardy to frost down to minus 12^0C \ fruits are eaten raw and cooked, flowers eaten raw and in preserves \ roots used to create a foam in beverages \ roots used as

soap substitute \ leaf fibre (40-50% dry weight) woven into cloth and mats and used as cordage \ during Second World War the fibre was used as a jute substitute, to make speciality paper and mattresses \ whole plant is chopped and mixed with cottonseed meal as a drought cattle fodder \ seed contains 29% semi-drying oil and 10% sarsasapogenin.

Yucca elephantipes

espinero
izote
palmito
yucca
Spanish dagger
spineless yucca

Central America \ frost tender, evergreen shrub to 10m tall \ swollen trunk at the base \ leaves to 1.2m long \ drought tolerant \ prefers well drained soils in an open sunny position \ not tolerant of extreme cold \ bitter ovaries and anthers are removed from flowers before the flowers are fried in batter \ they are also eaten in soups, stews and salads and are rich in vitamin C and niacin \ stem tips are eaten \ easily grown

from cuttings and when planted close together form an impenetrable barrier \ propagation from seed.

Yucca filamentosa

Adam's needle
Eve's thread
needle palm
common yucca
bear's thread

South western North America \ evergreen clumping plant to 5m tall \ leaves deep green, sword-shaped with white edges \ withstands saline, clay and dense soils \ best in full sun \ frost hardy \ plant flowers after about 5 years \ tall panicles of pendulous, tulip-shaped white flowers are eaten raw and said to be good in salads \ central spike eaten as a vegetable \ immature fruits are peeled then boiled and eaten \ edible, large, fleshy fruits called *datile* are eaten \ leaves a source of strong fibre and woven into baskets \ leaf fibre for cordage \ leaf point as a needle \ leaves and root contain nearly 2% saponin \ root was pounded for soap \ root salve or poultice to treat sprains and sores \ decoction of roots drunk to relieve arthritic pains \ plant used to treat headache, biliousness, depression, gland problems, gonorrhea, hepatitis, liver complaints, inflammation, irritability, melancholia, nervousness and rheumatism \ saponin from plant used to make soap, shampoo and cosmetics \ propagation is from seed but seedlings are very variable (for commercial purposes).

Yucca filamentosa

Yucca filifera

Mexico, where it is very common but being destroyed for agriculture \ slow growing - 30 years to produce seed \ alkaline sandy soils \ thrives in 350mm per annum rainfall areas \ leaves used locally as thatch and leaf fibre for making brushes and for paper pulp \ seeds rich in sarsasapogenin and contain 20% semi-drying oil (used as a plasticiser after epoxidation).

Yucca glauca
Yucca angustifolia
Yucca angustissima

small soapweed
soaproot
soapweed yucca
narrow leaf yucca
slender yucca
dwarf yucca
Spanish bayonet
plains yucca
Adam's needle

Tropical America \ short, clumping plant, without stems (to 0.6m tall and spreading to 1m) \ very hardy, often snow covered for weeks and able to withstand temperatures as low as minus 23^0C \ plant in full sun \ flower stems, flowers and fruit were all harvested from the wild for food \ fruit and flower buds were eaten fresh or roasted in the coals for about 15 minutes \ fruit eaten dried \ leaves were boiled with salt \ young spikes were eaten like cabbage \ saponin from plant extracted by crushing with stone is used to make soap, shampoo and cosmetics \ roots pounded and mixed with water to lather as a hair wash for treating dandruff, skin irritations (including poison oak rash) and as a strong laxative \ yucca root with

duck grease used by Hopi to cure baldness \ Apache used emulsion for snake and insect bites \ a cold infusion of the root used by the Ramah Navaho for prolonged labour \ plant used for fractures, saddle-sores, sprains, sores and as a haemostat \ leaf fibre for brushes \ baskets, bowstrings, G-strings and sandals from leaf fibre \ juice used as a varnish \ arrow poison from leaf juice mixed with charcoal from lightning-struck piñon or juniper.

Yucca gloriosa

Adam's needle
Spanish needle
Spanish dagger
mound lily

USA and north eastern Mexico \ stout stem, with a crown of tufted, long, sharp-pointed, deep-green, fleshy, relatively stiff leaves \ leaves to 800mm long and 50mm wide \ with a terminal spike of bell-shaped flowers \ best in full sun on well-drained soil \ tolerant of drought and very hardy to frost (to minus 5°C \

cultivated for edible flowers and fruit \ plant used as a bactericide, purgative, detergent and to treat asthma, bronchitis, dysentery, edema, hemorrhage, leprosy, phthisis, rheumatism, septicemia and sores \ roots poisonous \ propagation from seed or stem cutting.

Yucca harrimaniae
gobweb yucca
Utah \ young flower spikes eaten \ plant considered emergency food by southwest Indians.

Yucca macrocarpa
palma criolla
South west USA, Mexico \ edible fruit.

Yucca periculosa
Mexico \ petals used in chocolates.

Yucca schidigera
Yucca mohavensis
wild date
Mojave yucca
Spanish dagger
South western North America \ rarely exceeds 4.5m tall \lower reaches of many pinion-juniper woodlands, throughout desert shrub communities, and into coastal chaparral \ probably the most common yucca of the desert area \ flowers and fruits are eaten raw, roasted and in jellies \ young stems are prepared and eaten similar to sweet potato \ leaves pulped for soap \ leaf fibres were made into rope, twine, hats, hair brushes, shoes, mattresses, and saddle blankets

(experiments conducted during the early part of this century indicated that fibers derived from many yuccas could serve as a satisfactory substitute for jute because of their relatively high tensile strength and lustrous white appearance) \ extracts made from steroidal saponins of this plant are used to treat arthritis and as antistress agents for humans and poultry \ derivatives of these compounds are used as plant fertilizer, as additives to promote weight gain in cattle, and to lessen ammonia formation of poultry wastes \ is known to hybridise with Y. baccata and Y. constricta \ well adapted to survive most fires, usually sprouts from roots protected by overlying soil, or from surviving active tissues at the stem base.

Yucca schottii
amole yucca
schott yucca

South west USA, north western Mexico \ often branched tree to 6m \ leaves to 50mm wide and 800mm long with a sharp spine at the tip \ very hardy to frost down to minus 12^0C \ tolerant of shade \ fruit grows to 120mm long and

50mm wide is eaten raw, dried or cooked with flour

Yucca standleyi
South west North America \ fruit eaten roasted.

Yucca thompsoniana

Mexico \ grows to 4m \ fruit eaten.

Yucca torreyi
palma
Torrey yucca
South west North America \ grows to 9m \ single stem and long leaves \ very hardy to frost down to minus 12^0C \ fruit and young flower spikes eaten \ fruit baked then pounded and drained and the juice drunk or poured over cakes (below) \ roasted fruit pulp made into cakes which could be stored \ plant used to treat cough \ leaves used as main portion of baskets by Apache who also used roots to produce a red pattern in their baskets.

Yucca treculeana
Mexico and western Texas \ edible fruit is said to resemble a pawpaw \ plant used as a purgative.

Yucca whipplei
Hesperoyucca whipplei

quiote
chaparral yucca
Spanish bayonet
our Lord's candle

Mexico, California \ evergreen, stemless shrub \ slender, pointed, blue-green leaves form a dense tuft \ can grow to about 1m tall with flower stalks to 3m tall \ some forms die after seeding, others sucker to clumps and flower for years \ long flowering stem carries fragrant white flowers in panicles \ a single plant can bear six thousand blooms usually flowers after about 7 years \ rainfall is typically 250-350mm \ best in full sun on well-drained soil \ very hardy to frost down to minus 12^{0}C \ fruit and flowers eaten raw and in jellies \ seeds were ground and eaten as flour \ flowers eaten boiled \ stem centres and heads were eaten raw or roasted in pits by Amerindians \ fine fibre from leaves used as soft lining of saddle blankets \ a good barrier plant \ propagation from seed or sucker.

CACTACEAE

Most people associate cacti with desert conditions. Many of the cacti have evolved in deserts, some which do not receive rainfall for years at a time. Cacti have modified to cope under these extreme conditions. They have long root systems which seek out any water which they store, either in their roots or in their stems. Some from coastal Peru rely for years at a time on the fog that rolls across the landscape at night. Some have waxy coatings which reduces transpiration, others shade themselves in their own spines. Some cacti such as the giant saguaro cactus *(Carnegiea gigantea)* will only germinate in the shade. When overgrazing destroys the pioneers, in this case the palo verde tree *(Cercidium microphyllum, Cercidium floridum and Parkinsonia aculeata)*, the cacti no longer germinate. In the case of *Carnegiea gigantea* which lives for 300 years it may not be apparent that there are no young until it is too late.

Palo verde tree

There are also the rainforest dwelling cacti which do not need protection from the sun as they clamber up trunks and limbs of trees. Many of these are epiphytes and can live without contact with the soil. They can grow from a rotting trunk or in the crouch of branches where leaf litter accumulates.

Their uses to the people who have learned to live in the cactus domain are many and varied. All of the cacti come from the Americas although many have become naturalised in other parts of the world. Luther Burbank, the celebrated plant breeder, demonstrated the potential of the cactus by developing many different cultivars of *Opuntia ficus-indica* all without spines and each with a different fruit. Many of these are claimed to have been taken from stock selected by native people of different areas. Unfortunately these are now lost or so confused that they are generically referred to as 'Burbank spineless'. Some seedlings of the spineless prickly pear will throw back to thorn and will ultimately dominate a forage system through selective grazing by stock.

Amerindians made a poultice from seeds and blossoms of squash *(Cucurbita pepo)* and applied it to scratches from cactus thorns.

Many species are useful as emergency stockfeed, providing valuable food but also moisture for survival. Although the spines are not deadly to stock, they are generally cut off the larger species such as *Ferocactus* and burned off others but some are without spines and easily browsed.

Acanthocereus spp.
Dendrocereus spp.
Pseudoacanthocereus spp.
Tropical America, Caribbean to Brazil \ 11 species in genus.

Acanthocereus occidentalis
                      pitaya
Edible fruit.

Acanthocereus tetragonus
Acanthocereus acutangulus
Acanthocereus floridanus
Acanthocereus pitajaya
Acanthocereus pentagonus
Acanthocereus princeps
Cereus pentagonus

              pitahaya
             naranjada
     barbed wire cactus
                dildoe
         sword pear

Central America, West Indies, Mexico and southern USA (Florida) \ widespread, large, upright or creeping tree-like cactus to 9m tall, armoured with long spines \ the arching branches take root where they touch the ground \ prefers filtered light \ hardy to minus 12^{0}C \ night-blooming, greenish-white flowers are 200mm long, appear in late summer and produce edible red fruit with red flesh to 50mm long \ fruit eaten raw or cooked \ tender stems eaten cooked \ diuretic.

Ariocarpus spp.

Roseocactus spp.

Texas, New Mexico \ 7 species in genus \ all species at risk due to collectors and habitat destruction \ small, slow-growing, spineless, tubercled plants with deep turnip-like roots \ soil should comprise 50% perlite or pumice stone.

Ariocarpus agavoides

magueyitos
little agaves

Mexico / tubercles have a sweet taste and are eaten, often added to salad \ mucilage as a glue.

Ariocarpus fissuratus

Ariocarpus intermedius
Ariocarpus lloydii
Roseocactus fissuratus
Roseocactus intermedius
Roseocactus llodii

hikuli sunami
tsuwiri
sunami
living rock
peyote cimarrón
chaute

Southern Texas, northern Mexico \ slow-growing, flattened, globular cactus to 100mm tall and 150mm wide \ stem is grey and covered with warty, rough, triangular tubercles, each producing a tuft of wool \ limestone ridges \ large turnip-like root \ slow growing \ plant in full sun in a well drained soil \ plant often

survives exclusively on desert mists and dew \ semi-hardy to frost (-8^{0}C) \ juice used in preparing *tesguino*, an alcoholic drink made from mashed corn sprouts or corn stem juice \ used as a medicinal plant and as a pain killer, it is placed on bruises, wounds and bites \ used to relieve fevers and rheumatic pains \ it was chewed or drunk as a stimulant by runners \ a strongly intoxicating drink was made by boiling the plant in water for a few minutes \ flesh claimed to contain alkaloids similar to mescaline but discredited by Tarahumaris and Huichols tribes as stronger than peyote and evil, resulting in driving people mad \ mucilage as a glue \ propagation from seed.

Ariocarpus kotschoubeyanus

pezuna de venado
pata de venado

Mexico \ plant used externally on wounds \ mucilage as a glue.

Ariocarpus retusus

Ariocarpus furfuraceus

hikuli tsuwiri
tsuwiri
living rock
false peyote
seven stars
chaute

Mexico (vulnerable) \ slow-growing, small brown, purple-green or grey-green to 150mm in diameter, with short triangular tubercles \ hardy to frequent frosts as low as minus 8^{0}C \ requires full sun \ should be kept dry in winter \ plant used to treat fevers \ similar hallucinogenic properties to *Ariocarpus fissuratus* \ Huichols consider this a dangerous cactus, driving a person mad in the desert if they don't have guidance of a shaman \ mucilage as a glue.

Armatocereus spp.

Western South America \ 14 species in genus.

Armatocereus laetus

Lemaireocereus laetus

pishicol

Ecuador, Peru \ greyish-green columnar cactus \ 4-6m tall \ nocturnal white flowers produce edible fruit.

Browningia spp.

Peru, Bolivia, northern Chile \ 13 species in genus.

Browningia candelaris

Peru (rare), northern Chile (vulnerable) \ tree cactus with spineless trunk \ 5m tall and a trunk to 450mm diameter \ main trunk develops spines to 150mm long \ prefers sunny location \ temperature should not drop below 5^0C \ edible fruit is oval-shaped and about 75mm long \ grows from seed at 20-30^0C with 13-14 hours daylight per day.

Carnegiea gigantea
Cereus giganteus

saguaro
sahuaro
papago
giant cactus

South west USA, Mexico \ only member of genus \ the floral emblem of Arizona \ tree cactus to 18m tall and weighs many tons \ ribbed, columnar, branching and slow growing \ lives to about 300 years, first branches in 75 years \ tolerant of occasional frosts to minus 4^0C \ white, night-blooming flowers to 100mm \ edible red or purple fruit eaten raw or in preserves said to taste like fig with a hint of strawberry and peach \ fruit a staple food of Papago tribe who considered the *saguaro* harvest as the start of the new year (June-July) and was celebrated with an intoxicating, fermented drink made from the first fruits \ sun dried fruit made into large cakes and balls \ used to make candy \ fruit boiled without sugar to make preserves \ Apache made a drink from the fruit \ fruit used to make a syrup and the Papago and Maricopa tribes fermented this into an alcoholic beverage \ Apaches, Chiricahua and Mescalero used the fruit to flavour and sweeten intoxicating drinks \ seeds parched and stored to be later made into meal cakes \ seeds, rind and pulp are ground and used as flour and eaten in soups or made into a paste and used like butter \ edible oil from seeds \ seed as poultry feed \ stems contain a psycho-active compound called carnegine (also 5-hydroxycarnegine, norcarnegine, 3-methoxytramine) \ plant used to promote milk flow after childbirth \ Seri tribe used plant to treat rheumatism \ dead ribs used as splints for broken bones \ ribs were the major warp material for Papago basketry \ ribs used for roofing \ Seri made a caulking compound by mixing dried plant skeleton with sea lion oil \ burls used as cups and other containers \ ribs were made into drying racks for datil fruit and seeds \ woody skeletons used as timber for fences and *hogan* (hut) construction \ split ribs made into bird traps \ large ribs used for tools and handles and also split into two and used as wooden tongs for collecting cholla buds and joints \ ribs of dead plants also used for collecting saguaro fruit \ spines used as needles in tattooing \ propagation from seed in shade see front of this (CACTACEAE) section at 20-30^0C with 13-14 hours daylight per day.

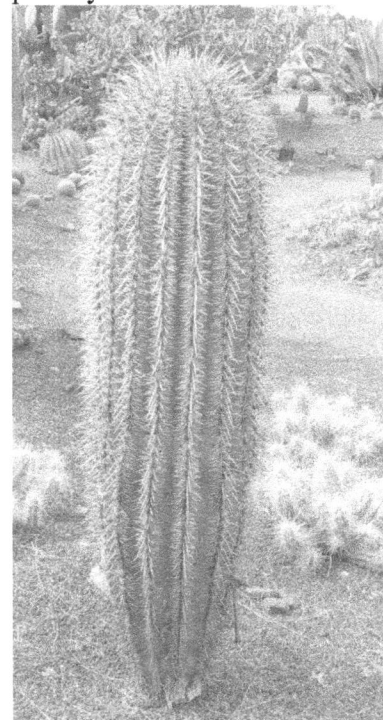

Cephalocereus spp.
Mexico \ 5 species in genus.

Cephalocereus columnae-trajani
Cephalocereus hoppenstedtii
Haseltonia columnae-trajani
Pachycereus columnae-trajani

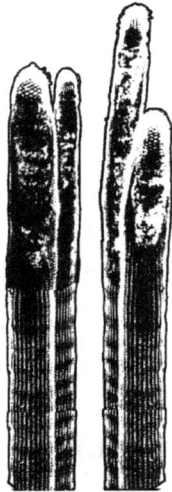

Oaxaca in Mexico \ large, columnar, unbranched cactus to 10m tall and 300mm diameter \ best in full sun on a calcareous soil \ nocturnal, white-yellow, summer flowers (at the top of the column) followed by edible fruit known locally as mountain peas \ strongly cathartic.

Cereus spp.
Caribbean and South America \ 63 species in genus \ tolerant of some frosts \ grows from seed at 20-30°C with 13-14 hours daylight per day.

Cereus aethiops
Cereus azureus
Piptanthocereus aethiops

Brazil, Uruguay, Argentina \ slender (40mm) to about 2m \ densely covered with black or dark-brown spines \ best in full sun \ tolerant of frequent frost to 18°C \ 200mm long pinkish-white flowers followed by edible fruit at a fairly early age.

Cereus bradtianus
Mexico \ edible fruit.

Cereus divaricactus
Plant used to treat warts.

Cereus donkelarii
Plant used as a cardiotonic.

Cereus fernambucensis
Cereus obtusus
Cereus pernambucensis
Cereus variabilis
Brazil to Uruguay \ narrow, oblong fruit is purplish red and splits open when ripe revealing an edible white pulp and black seeds.

Cereus fimbriatus
Plant used to treat warts.

Cereus hankeanus
Bolivia \ edible fruit.

Cereus hexagonus
clergé pascal
queen of the night
Caribbean, Surinam to Venezuela \ tree-like species to 15m \ edible, red, egg shaped fruit to 12.5cm in length \ stem parts as a vegetable \ plant used as a diuretic and to treat enterrhagia \ used as a hedge plant.

Cereus hildmannianus
Cereus alacriportanus
Cereus neonesioticus
Cereus peruvianus
Cereus uruguayanus
Cereus xanthocarpus
Piptanthocereus alacriportanus
Piptanthocereus neonesioticus
Piptanthocereus uruguayanus
Piptanthocereus xanthocarpus

apple cactus
Peruvian apple
pitaya
queen of the night

Brazil, Paraguay, Bolivia, Uruguay, Argentina \ columnar, erect, multiple-stemmed, ribbed, spiny, tree-like species to 16m \ stems have 5-8 ribs \ spines occur in groups of 7 \ occurs at medium elevation tropical climate \ best in full sun \ summer flowering \ nocturnal, waterlily-like, white flowers produce edible, globular, orange to red, oval fruit (to 5cm in diameter) with sweet, white, juicy, delicious flesh \ fruit peel cut into strips and candied \ younger stems eaten \ electromagnetic radiation protection \ can be cultivated as a hedge plant.

Cereus jamacaru
Cereus goiasensis
Piptanthocereus goiasensis
Brazil \ columnar, branched stem to 10m with 6 or more ribs \ yellow spines to 20mm long occur in clusters of 15 or more \ white flowers, tinged with green resemble water lilies and produce large, edible, bright red fruit \ edible stem used as a vegetable (stem also claimed to contain mescaline like San Pedro) \ electromagnetic radiation protection.

Cereus margaritensis
Venezuela to Colombia \ edible fruit.

Cereus quadrangularis
Plant used as a cancer treatment.

Cereus repandus
Cephalocereus repandus
Cereus grenadensis
Stenocereus peruvianus
Subpilocereus grenadensis
Subpilocereus remolinensis
Subpilocereus repandus

Caribbean, Venezuela \ large, columnar, tree cactus growing to 10m or more \ best in full sun \ night flowering in summer \ edible fruit \ stems peeled and eaten \ plant used to treat diarrhoea, as a soap and shampoo.

Cereus sp.
Monvillea cavendishii
Brazil, Argentina \ shrubby, clambering cactus \ edible fruit.

Cereus spegazzinii
Monvillea spegazzinii
Cereus anisitsii

Paraguay, Bolivia, Argentina \ shrubby, clambering cactus to 3m long and 15mm thick \ needs plenty of sun-light \ night flowering in mid-summer \ edible red fruit.

Cereus stenogonus
Piptanthocereus stenogonus
Paraguay, Bolivia \ edible fruit.

Cereus validus
Cereus forbesii

Argentina \ columnar cactus to 4m tall with erect branches \ full sun \ edible fruit.

Cleistocactus spp.
Borzicactus spp.
Northern Argentina, southern Brazil and western South America \ 49 species in genus \ fast-growing, columnar, much-ribbed stems and many spines \ species vary from very hardy to semi-hardy to frost \ require well-drained soil and full sun \ flowers are usually pollinated by hummingbirds \ grows from seed at 20-30^{0}C with 13-14 hours daylight per day.

Cleistocactus acanthurus

Borzicactus acanthurus
Loxanthocereus acanthurus
Loxanthocereus bicolor
Loxanthocereus brevispinus
Loxanthocereus canetensis
Loxanthocereus cantaensis
Loxanthocereus cullmannianus
Loxanthocereus eremiticus
Loxanthocereus erigens
Loxanthocereus eriotrichus
Loxanthocereus eulalianus
Loxanthocereus faustianus
Loxanthocereus gracilispinus
Loxanthocereus keller-badensis
Loxanthocereus multifloccosus
Loxanthocereus neglectus
Loxanthocereus pacaranensis
Loxanthocereus peculiaris
Loxanthocereus pullatus
Loxanthocereus xylorhizus
Borzicactus eriotrichus
Borzicactus gracilispinus
Loxanthocereus gracilispinus

Peru (vulnerable) \ low-growing, branching cactus with stems to 200mm long and 40mm in diameter \ does not tolerate frost and is best kept above 5^{0}C \ day flowering in summer \ edible fruit \ grows from seed at 20-30^{0}C with 13-14 hours daylight per day.

Cleistocactus baumannii

Cleistocactus aureispinus
Cleistocactus bruneispinus
Cleistocactus chacoanus
Cleistocactus flavispinus
Cleistocactus jugatiflorus
Cleistocactus santacruzensis

pitahayacita
scarlet bugler
Bolivia, Uruguay, Paraguay, Argentina \ erect, columnar cactus to 1m tall and 20-30mm diameter \ best in full sun \ red flowers appear in the day during summer and are followed by edible fruit.

Cleistocactus sepium

Cleistocactus jajoianus
Borzicactus aequatioralis,
Borzicactus sepium
Borzicactus ventimigliae
Borzicactus websterianus
Ecuador \ shrubby cactus of the Andes \ edible fruit.

Cleistocactus smaragdiflorus
Cleistocactus rojoi

sitiquira
Bolivia to Argentina \ erect, columnar cactus to 2m tall and 60mm diameter \ red-green, tubular flowers to 50mm long appear in summer during the day \ edible fruit \ *rojoi* is actually a hybrid with *Cleistocactus brookeae.*

Cleistocactus straussii
Southern Bolivia \ edible fruit

Coryphantha spp.
beehive cactus
South western USA and Mexico \ 63 species in genus \ grows from seed at 20-30^{0}C with 13-14 hours daylight per day \ remove from closed

humid environment a few days after germination.

Coryphantha compacta

doñana
wichuri
bakánawa
bakána
bakánori
híkuli
Santa Poli

Mexico \ small, solitary, slightly-flattened, globular, spiny cactus \ occurs naturally on dry, sandy hill and mountain country \ full sun \ very hardy to frost down to minus 12^{0}C \ used by Tarahuma shamans as a kind of peyote.

Coryphantha guerkeana
Coryphantha palmerii

doñana

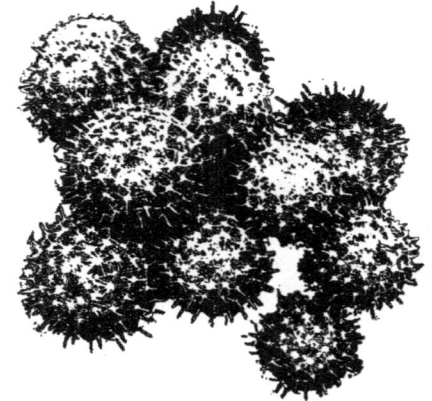

Durango in Mexico \ solitary or clumping, slightly elongated globular stems to 100mm diameter \ some frost tolerance to at least minus 4^{0}C \ lime soil preferred \ reported to have hallucinogenic properties.

Coryphantha macromeris
Coryphantha pirtlei
Lepidocoryphantha macromeris

doñana

Mexico, New Mexico, south west Texas \ low, clustering, elongated-globular cactus to 200mm with long, sharp spines \ branches at base covered with several spine tipped tubercules to 20mm long \ full sun \ very hardy to frost down to minus 12^0C \ spines are removed and 8-12 fresh or dried cacti are eaten on an empty stomach or crushed and brewed for an hour into a tea \ induces hallucinations in the user \ propagation from cutting or seed.

Disocactus spp.
Southern Mexico, Central America, Caribbean, north and western South America \ 22 species in genus.

Disocactus biformis
Cereus biformis
Phyllocactus biformis
Epiphyllum biforme

Honduras (rare) and Guatemala (rare) \ small, epiphytic cactus with flat, leaf-like, long branches to about 200mm long coming from a cylindrical stem \ acid, free-draining soil \ filtered light \ does not like cold \ day flowering in early spring, producing edible fruit.

Echinocactus spp.
Mexico and south western USA \ 6 species in genus \ barrel cactus \ grows from seed at 13-14 hours daylight per day.

Echinocactus equitans
Pulp from plant is sliced into strips and candied.

Echinocactus grusonii
golden barrel cactus
mother in law's cushion
golden ball

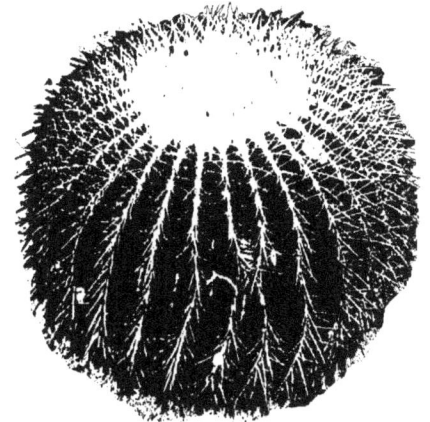

North America, Mexico (endangered)\ armed, spherical cactus with golden spines \ slow-growing to 1m tall and 1m wide \ eventually offsets to form clumps \ requires full sun in a well-drained position \ hardy to frequent moderate frost to minus 8^0C \ pulp used to make candy \ propagation from seed or cutting.

Echinocactus horizonthalonius
manca caballo
eagle's claw cactus
blue barrel cactus
mule crippler

South west USA, northern Mexico \ shrubby cactus to 250mm tall and 400mm wide \ 8 ribs \ medium elevation \ needs alkaline soil \ very hardy to frost down to minus 12^0C \ stem pulp used in confection and sweetmeat \ day flowers are rose or pink and occur in summer \ fruit sliced and candied.

Echinocactus platyacanthus
Echinocactus grandis
Echinocactus visnaga
Echinocactus ingens
Echinocactus karwinskii
Echinocactus palmeri
Melocactus ingens

biznaga
giant barrel
barrel cactus

Mexico (vulnerable) \ slow-growing, broad, globular cactus to 1m tall and 0.6m wide (early report of 2m tall and wide), with a grey-green stem with up to 50 ribs, 40mm sharp spines and a woolly crown \ hardy to frequent moderate frost to minus 8⁰C \ porous, calcareous soil \ full sun \ fleshy parts eaten in salads \ stem pulp for sweetmeat \ pulp of the stem is candied (*acitrón* or *cubiertos de biznaga*) and eaten as a sweet or as a flavouring \ potable drink from stems \ made into containers for food.

Echinocactus polycephalus
Echinocactus
xeranthemoides

water-storage parenchyma
devil's pin cushion

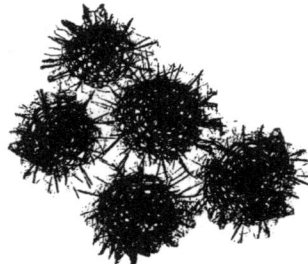

California, Nevada, Mexico (Sonora) \ globular to elongated cactus to 900mm tall and 250mm diameter \ solitary when juvenile, but forms large clumps with age \ very hardy to frost down to minus 12⁰C \ seed eaten by Panamint tribe of California \ berries and stems were a dependable and important food source of the Cahuilla \ Papago Indians cut the thorny rind away then after draining for a couple of days the pulp was cut into pieces and boiled in syrup from the fruit of *Carnegiea gigantea* to make a candy \ today the candy is usually made with sugar instead of fruit syrup \ spines as awls in basketry \ optimum seed germination at 17⁰C at night and 38⁰C daytime maximum \ keep reasonably dry after germination.

Echinocereus spp.
Mexico and south west USA \ 82 species in genus \ columnar to spherical cacti \ branch freely with age \ propagate from seed or stem cuttings \ optimum germination between 17⁰C at night and 20⁰C daytime maximum at 13-14 hours daylight per day \ remove from closed humid environment a few days after germination.

Echinocereus chisoensis
Echinocereus chiloensis
Echinocereus fobeanus
Echinocereus reichenbachii
var. chisoensis

lace cactus
South west USA (vulnerable), Mexico \ low cactus sometimes clumping to 0.1m tall and often singular to 0.2m tall \ very hardy to frost down to minus 12⁰C \ fleshy stem parts eaten as a vegetable.

Echinocereus cinerascens
Echinocereus
chloraphthalmus
Echinocereus ehrenbergii
Echinocereus glycimorphus
Echinocereus tulensis
Mexico \ low cactus to 0.3m tall and 70mm wide which clumps to a spread of about 1m \ frost tender below minus 5⁰C \ mature plants produce masses of bright pink or purple flowers up to 120mm across which finally produce edible fruit.

Echinocereus coccineus
Echinocereus arizonicus
Echinocereus neomexicanus

tjeenáyookísih
hedgehog cactus
crimson hedgehog cactus
sitting cactus
South west USA (vulnerable), Mexico \ low cactus to 1.5m tall \ green, warty stems form clumps with a spread of 1.5m \ full sun and well drained soil \ very hardy, often snow covered for weeks and able to withstand temperatures as low as minus 23⁰C \ funnel-shaped scarlet flowers produce edible fruits which were eaten by

many tribes \ Navajo used the cactus as a heart stimulant.

Echinocereus engelmannii
Cereus engelmannii
Echinocereus munzii
Indian strawberry
nichol hedgehog

South west USA (rare) and northern Mexico \ low cactus to 0.6m tall, forming clumps to about 1m diameter \ very hardy to frost down to minus 12^{0}C \ day flowering in summer \ spiny, pinkish-red, ovoid fruit is edible with a delicious strawberry flavour.

Echinocereus enneacanthus
Echinocereus dubius
Echinocereus merkeri
Echinocereus sarissophorus
Echinocereus uspenskii
Cereus macracanthus
strawberry cactus
hedgehog cactus
South west USA and Mexico \ clumping cactus to 200mm tall and 75mm diameter \ very hardy to frost down to minus 12^{0}C \ needs alkaline soil \ full sun \ flowers in mid-summer during the day \ edible red fruit very close in flavour to strawberry \ fruit drops its spines when ripe \ fruit popular in jam.

Echinocereus fendleri
Echinocereus abbeae
Echinocereus bonkerae
Echinocereus boyce-thompsonii
Echinocereus fasciculatus
Echinocereus hempelii
Echinocereus kuenzleri
Echinocereus ledingii
Echinocereus rectispinus
Echinocereus robustus
ho'ko
bonker hedgehog
hedgehog cactus

South west USA and Mexico \ low cactus to 0.2m tall, forming clumps to 1m wide \ full sun \ very hardy, often snow covered for weeks and able to withstand temperatures as low as minus 23^{0}C \ day flowering in summer \ ovoid, red fruit with whitish spines is edible \ stems eaten roasted \ dried fruit used by Hopi as a source of sweetening.

Echinocereus grandis

Islands in the Gulf of California \ low cactus to 0.3m tall and 100mm thick \ fleshy part used for candles.

Echinocereus pectinatus var. dasyacanthus
Echinocereus dasyacanthus
Echinocereus hildmannii
Echinocereus steereae
Cereus dasyacanthus
chihuahua
torch thistle
golden rainbow

South west USA and Mexico \ globose or columnar, low (150mm high and 80mm wide) cactus with green stem and white spines \ purple, pink or yellow, funnel-shaped flowers (to 120mm across) \ hardy to frost down to minus 6^{0}C \ full sun \ needs alkaline soil \ day flowering in summer \ the greenish-purple, gooseberry-like fruit is rich in sugar and said to be delicious.

Echinocereus polyacanthus
Echinocereus acifer
Echinocereus durangensis
Echinocereus leeanus
Echinocereus marksianus
Echinocereus pacificus
Echinocereus triglochidiatus var. acifer

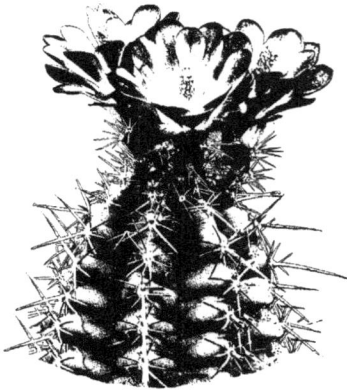

South west USA and Mexico \ wide distribution \ low mound-forming cactus \ long needle-like spines \ full sun \ very hardy to frost down to minus 12^0C, some provenances receive snow \ edible fruit \ stem used as false peyote by Tarahumara (contains tryptamines).

Echinocereus poselgeri
Cereus poselgeri
Wilcoxia kroenleinii
Wilcoxia poselgeri
Wilcoxia tamaulipensis
Mexico \ shrubby cactus with branching stems to 200mm long \ best in filtered sun \ summer flowering in summer \ edible fruit.

Echinocereus reichenbachii
Echinocereus caespitosus
Echinocereus mariae
Echinocereus perbellus
Echinocereus purpureus
tufted hedgehog cereus
South west USA and Mexico \ columnar, low cactus (150mm tall and 100mm wide) with spines about 15mm long \ very hardy, often snow covered for weeks and able to withstand temperatures as low as minus 23^0C \ trumpet-shaped, white, pink, red or purple flowers produce small purple edible fruit which are said to be delicious \ fleshy parts of the pads eaten as vegetable.

Echinocereus rigidissimus
Echinocereus pectinatus var. rigidissimus
Echinocereus pectinatus var. rubispinus
rainbow cactus

Mexico, USA \ cylindrical to globular cactus to 300mm tall and 80mm diameter \ full sun \ very hardy to frost down to minus 12^0C \ day flowering in early summer \ raw fruit eaten \ inner stem eaten and relished by Mexicans and Indians.

Echinocereus scheeri
Echinocereus cucumis
Echinocereus gentryi
Echinocereus ortegae
Echinocereus salm-dyckianus
Echinocereus salmianus

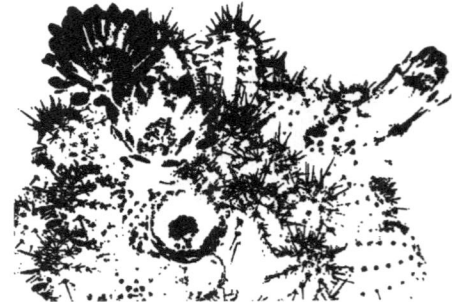

Mexico \ low, cylindrical, clumping cactus with yellow-green stems to 40mm in diameter and 150mm tall \ sometimes creeping habit \ full sun \ very hardy to frost down to minus 12^0C \ used as false peyote by Tarahumara and Huichol tribes.

Echinocereus stramineus
Echinocereus conglomeratus
alicoche
pitahaya de agosto
Mexican strawberry
strawberry cactus
South west USA and Mexico \ low cactus (to 200mm tall) forming massive clumps exceeding 1m wide \ best in sunny position on calcareous soil \ very hardy to frost down to minus 12^0C \ day flowering in mid-summer \ edible fruit with somewhat acid, strawberry flavour.

Echinocereus triglochidiatus
Echinocereus canyonensis
Echinocereus decumbens
Echinocereus gonacanthus
Echinocereus hexaedrus
Echinocereus krausei
Echinocereus kunzei
Echinocereus roemeri
Echinocereus rosei

pittallito
wichurí
claret cap
red hedgehog cactus

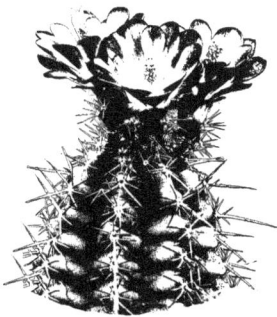

South west USA and Mexico \ low cactus to 0.15m tall in clusters up to 1.5m wide \ requires full sun \ flowering occurs in mid-summer during the day \ bright red, ovoid fruit to 25mm long is edible \ fruit pulp used like squash or made into sweet pickles \ used as a false peyote.

Echinopsis spp.
Trichocereus spp.
South America \ 154 species in genus.

Echinopsis bridgesii
Trichocereus bridgesii

achuma

Bolivia \ columnar, 4-8 ribbed, blue-green cactus, branching at the base \ full sun \ well drained soil \ reasonably frost hardy \ contains mescaline and said to make one drunk with visions \ contains more than 25mg per 100g fresh plant of mescaline \ grows from cutting or seed at 20-30^{0}C with 13-14 hours daylight per day.

Echinopsis chiloensis
Trichocereus chiloensis

cardón de candelabro
quisca
quisco

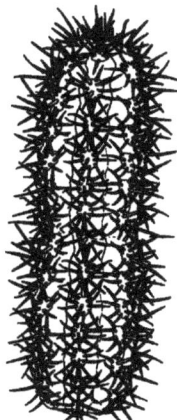

Chile to Argentina \ large, columnar cactus to 7m tall and 120mm diameter \ best in full sun \ night blooming in summer \ sweet, mucilaginous, edible

fruit eaten raw or in syrup and brandy \ plant used to treat tumours \ a raw material, called *normata*, which is dry sections of this plants, is gathered on the ground and then used in workshops to manufacture a product, known as *palo de agua, palo de lluvia* or *palo musical* \ grows from seed at 20-30^{0}C with 13-14 hours daylight per day.

Echinopsis cuzcoensis
Trichocereus cuzcoensis

upright organ-pipe

Peru \ rocky slopes from 2500-3000m altitude \ stem contains hallucinogenic mescaline \ plant used to treat cancer \ small-scale workshops in Peru manufacture the product *palo de lluvia* from the stem \ grows from seed at 20-30^{0}C with 13-14 hours daylight per day.

Echinopsis deserticola
Echinopsis fulvilana
Trichocereus deserticola
Trichocereus fulvinanus
Chile (vulnerable) \ columnar cactus to 1.5m tall and 75mm diameter \ best in full sun \ stem contains hallucinogenic mescaline \ grows from seed at 20-30^{0}C with 13-14 hours daylight per day.

Echinopsis macrogona
Trichocereus macrogonus
Bolivia \ stem contains hallucinogenic mescaline\ grows from seed at 20-30^{0}C with 13-14 hours daylight per day.

Echinopsis oxygona
Echinopsis multiplex
Trichocereus oxygona

Brazil, Uruguay, Argentina \ globular plant to about 150mm tall \ forms off-sets from the base \ full sun \ stem contains hallucinogenic mescaline \ grows from seed at 20-30^{0}C with 13-14 hours daylight per day.

Echinopsis pachanoi
Echinopsis peruviana
Trichocereus pachanoi
Trichocereus peruvianus

San Pedro
aguacolla
cardon
cimarrón
cimora blanca
cuchuma
gigantón
huando hermosa
San Pedrillo
símora
cactus of the four winds
huachuma
achuma
Peruvian torch

Peru, Ecuador, Bolivia \ tall, columnar, branching cactus to 6m \ reddish-brown night blooming flowers to 250mm long \ naturally occurs at altitudes 1800-3000m \ best in sandy or well-drained soil, with plenty of sun \ large white flowers \ grown for edible fruit said to be sweeter than prickly pear \ plant contains 0.01% mescaline (about a tenth of peyote) the hallucinogenic alkaloid (*var. peruvianus* is claimed to contain at least 10 times the quantity, perhaps as testament to Andean gardeners ability to selectively breed plants) \ the plant was the centre of religious rituals amongst Andean Indians for at least 3000 years \ the Spanish tried to stop the practice with little success \ one piece of cactus 75mm wide and 30-150mm long is considered a dose \ it can be peeled, cut and eaten, cut into slices and dried or mashed and boiled in a litre of water for 2 hours then strained and sipped slowly over about 45 minutes \ the inner green pulp next to the outer skin is believed to be the part used \ plant used as an emetic and to treat enteritis, gastritis, pneumonia and sterility \ small-scale workshops in Peru manufacture the product *palo de lluvia from the stem* \ very popular rootstock for grafting other cacti species onto \ easily grown from cutting or seed.

var. peruvianus on left compared to the thorny specimen on the right grown from seed which was collected from the wild.

Echinopsis pasacana
Echinopsis rivierei
Cereus pasacana
Helianthocereus pasacana
Leucostele rivierei
Trichocereus pasacana
Trichocereus rivierei

cardón santos

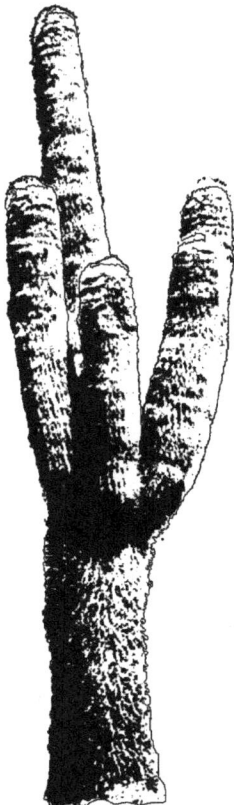

Bolivia (rare) to northern Argentina \ tall tree cactus to 5m tall and 500mm wide at the base \ requires full sun \ day flowering in summer \ edible, round, green fruit \ trunks used for huts and fencing \ flower and fruit ash used as a lime substitute for treating coca (*Erythroxylum species*).

Echinopsis schickendantzii
Echinopsis manguinii
Trichocereus manguinii
Trichocereus shaferi
Trichocereus schickendantzii

Argentina \ cactus to 500mm tall and 120mm wide \ best in full sun \ day flowering in summer \ fruit eaten raw or made into brandy and syrup.

Echinopsis spachiana
Echinopsis santiaguensis
Cereus spachianus
Trichocereus santiaguensis
Trichocereus spachianus

torch cactus
golden torch
desert cactus

Argentina, Bolivia \ clumping, ribbed, cylindrical, erect,

columnar stems (about 75 mm thick) branching at the base and rising to about 7m \ 10-15 broad ribs \ pale golden spines \ requires full sun and well drained soil \ frost tender (hardy to 5^{0}C) \ drought hardy \ nocturnal, fragrant, funnel-shaped, white flowers to 200mm long and 100mm in diameter appear in summer \ edible fruit eaten raw or used in sherbets and ice cream \ new tender shoots added to vegetable salad or boiled and served with butter, pepper and salt.

Echinopsis taquimbalensis
Trichocereus taquimbalensis
Bolivia \ stem contains hallucinogenic mescaline.

Echinopsis terscheckii
Echinopsis werdermannian
Trichocereus culpensis
Trichocereus terscheckii
Trichocereus werdermannian
Argentine saguaro
Cardón santos (holy cactus)
Cardon grande
Bolivia, Argentina \ columnar plant to 12m tall and branching from or above the base \ branches about 150mm diameter \ best in full sun \ stem contains hallucinogenic mescaline (5 to 25mg per 100grams fresh plant) \ used to make coca taste better and give it increased strength.

Echinopsis tunariensis
Trichocereus tunariensis
Bolivia \ large cactus \ edible fruit.

Echinopsis valida
Trichocereus validus
Paraguay \ stem contains hallucinogenic mescaline (over 25mg per 100 grams fresh plant).

Epiphyllum spp.
Phyllocactus spp.
Caribbean, tropical America \ 19 species in genus.

Epiphyllum cooperi
orchid cactus
Whole fruit eaten peeled.

Epiphyllum lepidocarpum
Costa Rica \ bushy epiphyte \ edible flowers.

Epiphyllum oxypetalum
Epiphyllum latifrons
Cereus oxypetalus
Phyllocactus oxypetalum
keng hua
Dutchman's pipe

Mexico, Central America \ bushy epiphyte to 2m or more \ cylindrical stems have flat, notched, sprawling leaf-like branches \ best in humid, shaded conditions \ frost tender \ large, white, nocturnal flowers resemble waterlilies and produce edible fruits.

Epiphyllum phyllanthus var. hookeri
Epiphyllum hookeri
Epiphyllum stenopetalum
Epiphyllum strictum
Phyllocactus hookeri
Phyllocactus phyllanthus
caraguala
rabo de iguana
reina del baile

Mexico to Peru \ bushy epiphyte to 1m \ main stem is cylindrical with thin, flat, leaf-like, serrated branches up to 75mm wide \ frost tender \ fruit is edible \ edible flowers are white, funnel shaped and nocturnal, appearing in mid-summer \ flowers are used as a cardiac tonic.

Epiphyllum rosetta
Fruit eaten.

Epithelantha spp.
South western USA, north eastern Mexico \ 7 species in genus.

Epithelantha micromeris
Mammillaria micromeris
hikuli mulato
button cactus

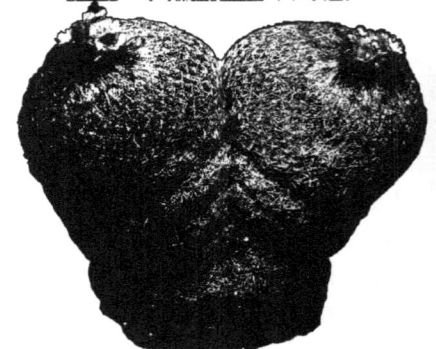

Mexico, Texas \ small, clumping or singular, usually globular, button cactus to 60mm in diameter with dense, white spines \ usually clumps with age \ alkaline soil essential for good health \ very hardy to frost down to minus 12^0C \ day flowering in summer \ edible, club-shaped, acidic fruit called *chilitos* grow to 13mm long and have large black shiny seeds to 2mm across \ Tarahumara tribe of Chihuahua use the plant as a false peyote to make their sight clearer, allow them to commune with sorcerers and act as a stimulant for runners \ propagation from seed which germinates readily but seedlings have low survival rate \ grows from seed at 20-30^{0}C with 13-14 hours daylight per day \ remove from closed humid environment a few days after germination.

Escobaria spp.

Western USA, northern Mexico \ 31 species in genus \ small spiny plants \ some frost tolerance \ grows from seed at 20-30^{0}C with 13-14 hours daylight per day \ remove from closed humid environment a few days after germination

Escobaria missouriensis

Escobaria asperispina
Neobesseya missouriensis
Neobesseya notesteinii
Neobesseya rosiflora
Neobesseya similis
Neobesseya wissmannii

Missouri pincushion

South west and central USA \ low-growing, solitary or clumping, globular, spiny cactus to 150mm wide \ very hardy, some provenances often

snow covered for weeks and able to withstand temperatures as low as minus 23^{0}C \ full sun \ day flowering in summer \ bright red, spherical (to 10mm diameter), edible fruit.

Escobaria vivipara

Escobaria aggregata
Escobaria arizonica
Escobaria bisbeeana
Escobaria chlorantha
Escobaria oklahomensis
Escobaria radiosa
Coryphantha aggregata
Coryphantha alversonii
Coryphantha arizonica
Coryphantha columnaris
Coryphantha vivipara

sour cactus
bisbee beehive
rose beehive

Alberta in Canada to southern USA, in a wide belt \ very variable but normally solitary, low cactus (50mm) having a green stem covered densely with grey spines \ very hardy,

some provenances often snow covered for weeks and able to withstand temperatures as low as minus 23^{0}C \ best in well-drained sunny position \ best kept dry in winter \ flowers in summer during the day \ edible, egg-shaped, red fruit about 3cm long is said to be pleasantly sour \ Navajo sun dried fruit and used them like currants.

Escontria chiotilla

chiotilla

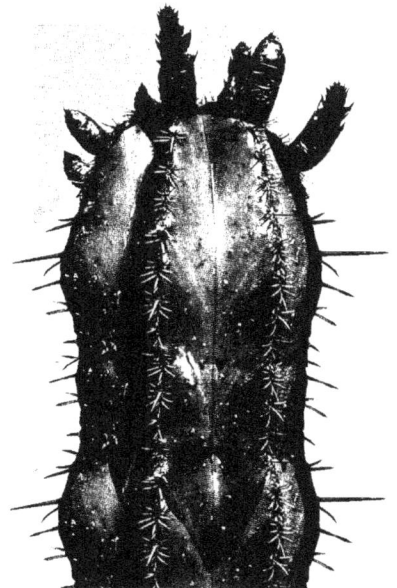

Mexico \ only species in genus \ branching, columnar tree cactus to 7m tall and a trunk to 400mm diameter \ full sun \ day flowering in mid summer \ fruit (*geotilla* or *tuna*) is purple, fleshy and very edible \ dried fruit with a flavour like gooseberry \ fruit used for ice cream and marmalade.

Espostoa spp.

Western South America \ 16 species in genus.

Espostoa lanata
Espostoa dautwitzii
Espostoa laticornua
Espostoa procera

Peruvian old man
pichcol negro
cottonball

Ecuador, Peru \ woolly, columnar tree cactus to 6m tall and 150mm diameter \ tolerant of occasional low frost above minus 4^{0}C \ foul-smelling, white flowers appear in summer and produce slightly oblong, smooth, red, sweet, berry-like fruit (to 60mm in diameter) is edible and called "*piscol colorado*" or "soroco".

Eulychnia spp.
Southern Peru, Chile \ 7 species in genus.

Eulychnia acida
Cereus acidus

copao
ácido

copao in use as a fence
Alacama desert in Chile \ woolly, slow growing, columnar cactus to 4m tall \ very low rainfall and well-drained soil \ edible, fleshy fruit is fairly acidic \ a raw material, called *normata*, which is dry sections of this plants, is gathered on the ground and then used in workshops to manufacture a product, known as *palo de agua, palo de lluvia* or *palo musical* \ used for living fences.

Ferocactus spp.
barrel cactus
South western USA, Mexico \ 35 species in genus \ genus of slow-growing, spherical cacti which become columnar after many years \ flower buds of many species are eaten boiled or fried after being soaked in water overnight and are said to taste like artichokes \ grows from seed at 20-30^{0}C with 13-14 hours daylight per day \ remove from closed, humid environment a few days after germination.

Ferocactus coulteri
barrel cactus
Seri tribe harvested drinking water where necessary.

Ferocactus cylindraceus
Ferocactus acanthodes
Ferocactus rostii
Echinocactus acanthodes

barrel cactus
compass barrel
desert barrel cactus
Californian barrel
fire barrel

South west North America, northern Mexico \ slow-growing, columnar, solitary (occasionally offsetting), sturdy, cylindrical, spiny cactus to 3m tall and 0.8m wide \ 8 ribs \ low to medium elevation \ requires full sun and well drained soil \ can tolerate frost in dry season down to minus 6^{0}C \ yellow, funnel-shaped flowers form in summer during the day and result in egg-shaped, yellowish-green fruit to 4cm long \ fruit pulp made into a syrup, eaten raw, dried or in preserves and candies \ top cut off plant and stem hollowed out flesh pounded to produce a potable liquid in times of thirst \ barrel can then be used as a cooking pot by filling with food to be cooked, adding heated rocks from the fire and stirring to avoid walls of pot from burning through \ seed ground as a flour for cakes \ flower buds eaten fresh, cooked or sun-dried for storage \ flesh is fed to stock as emergency feed \ spines set in pitch and used as awls \ spines used as tattooing needles.

Ferocactus emoryi

Ferocactus covillei
Ferocactus rectispinus
Echinocactus emoryi

barrel cactus

Arizona, Mexico \ globular becoming cylindrical to 1.5m tall and 600mm diameter \ full sun \ calcareous, porous soil \ tolerant of occasional low frost above minus 4^{0}C \ flesh is cooked in sugar and made into candy \ source of emergency drinking water \ seed eaten.

Ferocactus hamatacanthus

Hamatocactus
hamatacanthus
Hamatocactus sinuatus

Turk's head
lemon cactus

Texas, northern Mexico \ very slow-growing, ribbed, spherical to columnar cactus to 600mm tall and 300mm wide \ very hardy to frost down to minus 12^{0}C \ full sun \ day flowering in summer \ thin-skinned, acid-flavoured, juicy, greenish fruit

to 50mm stays on plant all winter and is used as a lemon substitute \ dried fruit eaten as a sweetmeat \ unopened flower buds soaked in water over night, boiled or fried, then eaten, tasting like artichokes.

Ferocactus histrix

Ferocactus electracanthus
Ferocactus melocactiformis
Echinocactus histrix

biznaga
Mexican barrel cactus
barrel cactus
melon cactus

Mexico \ solitary, spherical, spiny cactus elongating to 700mm tall \ low to medium elevation \ best in full sun on slightly calcareous soil \ hardy to frost down to minus 10^{0}C \ flowers during the day in mid-summer \ spherical, reddish fruit grows to 25mm in diameter and is eaten raw \ fruit juice made into wine \ flesh candied \ stem pulp yields a potable water \ made into containers for food.

Ferocactus viridescens

Ferocactus californicus
Ferocactus orcuttii
Echinocactus viridescens

coast barrel cactus

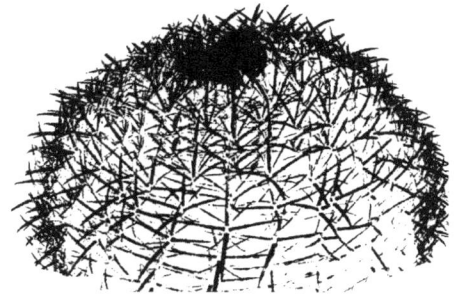

South west North America, Mexico (vulnerable) \ solitary, spherical cactus to 0.5m tall and 0.35m wide \ commonly offsets from the base \ naturally occurs at low elevation \ hardy to frequent moderate frost to minus 8^{0}C \ yellowish-green flowers form in the day in summer \ reddish, egg-shaped, fruit to 20mm long is edible, slightly acid and said to taste like gooseberry \ flower buds are cooked and eaten \ black seed eaten.

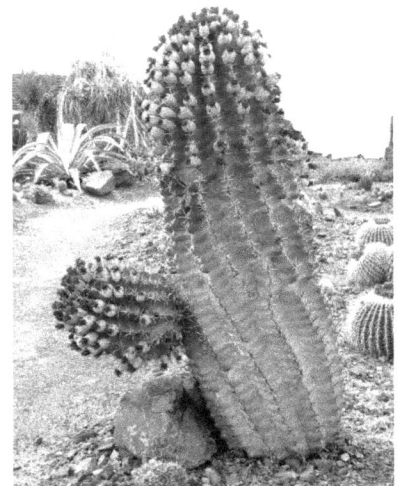

Ferocactus wislizenii
Ferocactus arizonicus
Ferocactus falconeri
Ferocactus phoeniceus
Echinocactus wislizenii

bisnaga
fish hook barrel
barrel cactus

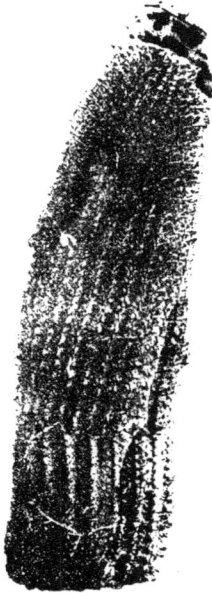

Mexico, California \ slow-growing, thorny, solitary, spherical becoming cylindrical cactus to 3m tall and 0.6m thick \ hardy to frequent moderate frost to minus 8^0C in dry season \ mostly occurs at medium elevation \ top cut off plant and stem hollowed out then flesh pounded to produce a potable liquid in times of thirst \ barrel can then be used as a cooking pot by filling with food to be cooked, adding heated rocks from the fire and stirring to avoid walls of pot from burning through \ pulp eaten as greens in summer \ yellow or orange, funnel-shaped flowers to 60mm across appear in the day in summer and are eaten toasted \ yellow fruit to 50mm long is sour and eaten \ stems sliced into small pieces with mesquite *(Prosopis sp.)* beans and eaten as a sweet dish by Pima \ stems candied in sugar \ parched seed ground for bread or gruel \ flower buds *(cabuches)* are pickled as an appetiser \ candles made from stem pulp \ spines heated and bent for use as fish hooks \ Seri used hollowed plant as a storage vessel for honey and the candied walls were then eaten by the children.

Gymnocalycium spp.
Twenty six species of this genus have been found to contain mescaline. The ones with the greatest concentration were: *Gymnocalycium calochlorum, Gymnocalycium comarapense, Gymnocalycium horridispinum, Gymnocalycium netrelianum, Gymnocalycium riograndense, Gymnocalycium striglianum, Gymnocalycium uebelmannianum, Gymnocalycium valnicekianum* and *Gymnocalycium vatteri.*

Haageocereus spp.
Bolivia, Peru, Northern Chile \ 19 species in genus \ best in full sun \ does not tolerate frost and is best kept above 5^0C.

Haageocereus decumbens
Haageocereus ambiguus
Haageocereus litoralis
Haageocereus mamillatus
Borzicactus decumbens

Peru, Chile \ edible fruit.

Haageocereus multangularis var. pseudomelanostele
Haageocereus acanthocladus
Haageocereus akersii
Haageocereus aureispinus
Haageocereus chosicensis
Haageocereus chrysacanthus
Haageocereus clavatus
Haageocereus crassiareolatus
Haageocereus dichromus
Haageocereus divaricatispinus
Haageocereus longiareolatus
Haageocereus pachystele
Haageocereus piliger
Haageocereus pseudomelanostele
Haageocereus setosus
Haageocereus symmetros
Haageocereus tenuispinus
Haageocereus turbidus
Haageocereus viridiflorus
Haageocereus zehnderi

Peru \ columnar cactus to 700mm tall \ full sun \ nocturnal flowers in summer \ edible fruit.

Harrisia spp.
Eriocereus spp.
Florida, Caribbean, Brazil, Bolivia, Paraguay, Argentina \ 23 species in genus.

Harrisia aboriginum
Cereus gracilis, var. aboriginum
shellmound apple-cactus
South east USA \ shrubby cactus \ yellow fruit eaten raw.

Harrisia eriophora
Cereus eriophorus
Grant apple cactus
Cuba \ shrubby cactus \ edible fruit.

Harrisia fragrans
Cereus eriophorus, var. fragrans
fragrant apple cactus
South east USA (endangered)\ shrubby cactus \ red edible fruit is eaten raw.

Harrisia gracilis
Harrisia donae-antoniae
Cereus gracilis

Jamaica, USA \ sprawling cactus to 4m long and 30mm diameter \ best in semi-shade \ nocturnal flowers in late summer \ edible fruit.

Harrisia guelichii
Eriocereus guelichii

Argentina \ slender-stemmed, creeping cactus to 30mm thick \ sunny site \ edible fruit.

Harrisia martinii
Eriocereus martinii
Cereus martinii

**Harrisia cactus
moonlight cactus**

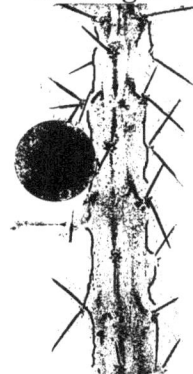

Argentina and Paraguay \ vigorous, creeping succulent \ best emerging from the shade of trees \ In Queensland, Australia it naturalised in the brigalow *(Acacia harpophylla)* \ large, white, funnel-shaped night-blooming flower

produces an edible red fruit with white flesh and many black seeds \ cultivated as source of alkaloid.

Harrisia pomanensis
Harrisia bonplandii
Cereus bonplandii
Eriocereus bonplandii
Eriocereus pomanensis

Brazil to Argentina \ slender-stemmed, creeping succulent to 3m in length \ good in partial shade \ tolerant of occasional low frost above minus 4^0C \ nocturnal flowers are 250mm long and appear in summer \ large, edible, red, scaled fruit \ grow from cutting or from seed at 20-30^{0}C with 13-14 hours daylight per day.

Harrisia simpsonii
Harrisia deeringii
Cereus gracilis var.
simpsonii

prickly apple
apple cactus

South east North America \
grows into mangrove swamps \
shrubby or erect cactus to 7m \
stems often vine like \ can be
terrestrial or epiphytic \ juicy,
orange-red fruit grows to
60mm and is eaten raw.

Harrisia tortuosa
Harrisia arendtii
Cereus tortuosus
Eriocereus arendtii
Eriocereus tortuosus

Harrisia cactus

Bolivia, Paraguay, Uruguay,
Argentina \ well armed,
creeping succulent capable of
forming impenetrable barriers \
does not tolerate frost and is
best kept above 5^0C \ edible
fruit \ roots eaten cooked.

Heliocereus spp.
Mexico, Guatemala \ 4 species
in genus.

Heliocereus speciosus
Heliocereus amecamensis
Heliocereus serratus
Heliocereus speciosissimus
Heliocereus superbus
Cactus speciosus
Disocactus speciosus

sun cactus

Mexico \ freely branching,
erect or clambering, epiphytic
cactus to 1m in length with
bright green stems with 3 or 4
narrow ribs \ partial shade \
slightly acidic, rich soil \ not
tolerant of frost \ edible, spiny,
scarlet, funnel-shaped flowers
produce red, egg-shaped fruit
to 50mm in length.

Heliocereus X Disocactus hybrid
Epiphyllum ackermannii
Nopalxochia ackermannii

red orchid cactus

Mexico \ naturally occurring
hybrid \ epiphytic, erect,
becoming pendant, fleshy-
toothed, green-stemmed cactus
to 300mm high with a 600mm
spread \ 150mm wide, red,
funnel-shaped flowers produce
edible fruit.

Hylocereus spp.
Mexico, Central America,
Caribbean, northern South
America \ 30 species in genus \
climbing, epiphytic cacti \ does
not tolerate frost and is best
kept above 5^0C \ grows from
seed at 20-30^{0}C with 13-14
hours daylight per day.

Hylocereus costaricensis
pitaya
Costa Rica, Nicaragua,
Panama \ clambering, epiphytic
cactus \ large, red, edible, pear-
shaped fruit.

Hylocereus guatemalensis
Cereus guatemalensis
Cereus trigonus var.
guatemalensis

pitaya

Guatemala, El Salvador \
clambering, epiphytic cactus
from the forest \ three-angled
stems grow to 5m long and
75mm wide \ filtered sun \
slightly acid soil \ night
flowering in summer \ edible
red-fleshed fruit.

Hylocereus ocamponis
Cereus ocamponis

pitahaya roja

Mexico \ clambering, epiphytic cactus of the forest \ three-angled, stems to about 3m long and 70mm wide \ requires partial shade \ nocturnal flowers in mid summer \ edible, red-wine coloured fruit is sweet and eaten raw \ often used as a hedge plant.

Hylocereus polyrhizus

pitahaya

Panama to Colombia \ clambering, epiphytic cactus \ edible red, thick skinned fruit with a sweet-flavoured, white or pink flesh \ juice used in making *refrescos* \ used in Ecuador as a hedge plant.

Hylocereus triangularis
Hylocereus compressus
Hylocereus cubensis
Cereus triangularis
Mediocactus coccineus

yellow pitaya
katom yellow

Cuba, Jamaica, Dominican Republic \ clambering, epiphytic cactus \ edible fruit \ in Colombia two annual harvests produced 6 tons of fruit per hectare from five year old plantation.

Hylocereus trigonus
Hylocereus antiguensis
Hylocereus napoleonis
Hylocereus plumieri
Cereus napoleonis

Puerto Rico, Virgin Islands, St. Vincent, Lesser Antilles \ clambering, epiphytic cactus with three-angled winged stems \ prefers good sun \ enjoys humid, warm conditions \ day flowering in mid-summer \ edible fruit.

Hylocereus undatus
Hylocereus tricostatus
Cereus undatus

pitahaya
pitahaya blanca
pitahaya roja
pitahaya de cardón
pitahaya oregona
wild pitaya
strawberry pear
dragon fruit
dragon pearl fruit
thang loy
queen of the night
night blooming cereus
Honolulu Queen

Origin unknown, common in cultivation since pre-history \ freely branching, clambering, epiphytic or terrestrial, three-sided cactus to 6m in length and 60mm across \ it may cling to branches, rocks and walls with strong aerial roots \ tropical climate \ night blooming flowers up to 350mm long and 225mm wide in summer \ unopened flower buds are eaten cooked \ hand pollination recommended for maximum fruit set \ juicy bright red fruits to 125mm in length are sweet and refreshing, eaten raw or made into preserves and sherbets \ stem juice used as a vermicide.

Lepismium spp.
Brazil, Bolivia, Argentina \ 15 species in genus.

Lepismium aculeatum
Rhipsalis aculeata
Brazil, Argentina, but naturalised in West Africa \ dry areas \ climbing, freely-branching, epiphytic succulent of closed forest \ edible, yellowish-brown, berry-like fruit.

Lepismium ianthothele

Lepismium erectum
Lepismium mataralense
Pfeiffera erecta
Pfeiffera gracilis
Pfeiffera ianthothele
Pfeiffera mataralensis
Pfeiffera cereiformis
Cereus ianthothelus
Cereus ianthothele
Rhipsalis cereiformis
Rhipsalis ianthothele
Hariota cereiformis

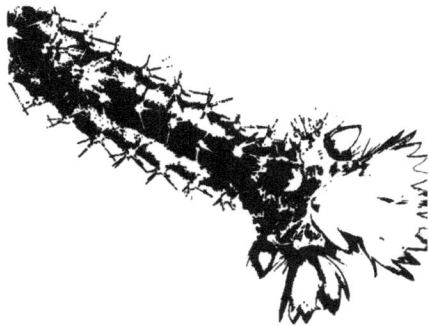

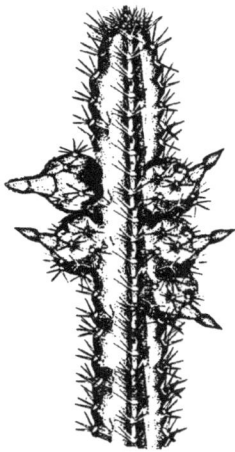

Bolivia to Argentina \ erect to pendant (both forms shown), epiphytic succulent to 0.5m long and 20mm wide with quadrangular stem \ grows in partial shade of jungle and rain forest \ grow best in humus-rich, moist compost, receiving only filtered light \ does not tolerate frost and is best kept above 5^0C \ white, cup-shaped flowers (30mm across) occur in the day in early summer \ pale, purple, spherical (15mm diameter), rosy-red, spiny, mistletoe-like berries which are edible \ propagate from cuttings.

Lophophora spp.

Texas, Mexico \ 2 species in genus.

Lophophora diffusa

Lophophora lutea

peyote de querétaro

Driest and stoniest deserts of Mexico (rare and vulnerable) \ grey-green or yellow-green crown and similar in appearance to and often considered synonymous with *Lophophora williamsii* \ sunny site \ slightly calcareous well-drained soil \ similar uses although claimed not to be as strong.

Lophophora williamsii

Lophophora echinata
Lophophora fricii
Lophophora jourdaniana

peyote
peyotl
peyotillo
hikuli
mescaline
mescal button
dumpling cactus
brandy head
raiz diabólica

Central Mexico to south Texas \ spineless, slow-growing, flat, globular stems to 80mm wide and about 60mm tall \ small pink flowers become pink berries \ requires well drained, slightly calcareous soil and full sun \ semi-hardy to frost \ hottest, driest, stoniest of deserts \ sandy, clay, lime soils under light shade \ Amerindians harvest the cactus tops (not the roots) which are called buttons \ buttons contain some 60 different alkaloids including: mescaline, and anhalonin and are used as an hallucinogen in many Indian religious ceremonies \ although the root also contains the psycho-active alkaloids, the plant survives if the root is spared and produces a cluster of buttons which can be later harvested \ a well managed clump of button clusters can reach 1.5m across \ westerners in search of the plant have not observed the same respect for the plant and consequently it is threatened with extinction in many areas of its natural range \ propagation from seed and division \ can also be grafted on to other root stock such as *Cereus peruvianus* to provide faster growth and better water tolerance in cold conditions \ propagation can be from cuttings of peytleros (young buttons) which grow from harvested roots.

Mammillaria spp.
pincushion cactus

USA to Venezuela \ 391 species in this diverse genus of columnar, spherical or hemispherical cacti \ generally require full sun on a well-drained soil \ some species will tolerate partial shade \ grows from seed at 13-14 hours daylight per day \ one species known as ball cactus or Navaho testicles was treated by burning the spines off, then the entire plant was eaten raw.

Mammillaria craigii
whichuriki
hikuli rosapara
hikuri
peyote de San Pedro

Mexico \ globose with woolly, white spines and rose coloured flowers to 15mm long \ used as a medicine for relief of headaches, earaches and deafness \ a false peyote of the Tarahumara tribe.

Mammillaria dioica
Mammillaria estebanensis
fish hook cactus

USA, Mexico \ small fruit eaten raw \ does not tolerate frost and is best kept above 5^0C \ optimum germination between 20^0C at night and 30^0C daytime maximum \ remove from closed, humid environment a few days after germination.

Mammillaria fissurata
dry whisky

Chewing causes intoxication.

Mammillaria grahamii
Mammillaria microcarpa var. grahamii

whichuriki
hikuli rosapara
hikuri
peyote de San Pedro
fish hook cactus
fish hook pincushion
snowball pincushion
sunset cactus

USA \ (rarely clumping) globose or cylindrical cactus to 100mm in diameter \ full sun \ does not tolerate frost and is best kept above 5^0C \ day flowering in summer \ small fruit eaten raw or dried and cooked \ Pima boiled the plant and placed a warm portion in the ear to treat ear ache and

suppurating ears \ plant, including fruit, used as false peyote by Tarahumara.

Mammillaria heyderi
Mammillaria gaumeri

coral cactus
nipple cactus
biznaga de chilillos
híkulu

South west USA, north eastern Mexico \ single or multi-headed, small, globular or slightly cylindrical, spiny cactus to 110mm high and clumps can reach 300mm wide \ medium elevation \ different provenances of this species range from no frost tolerance to very frost tolerant \ full sun \ day flowering in summer \ bright red, elongated, fruit (chiitos) grows to 25mm long and is edible \ piece of plant is roasted for a few minutes and then the centre is pushed into the ear to treat earache or deafness \ plant used as false peyote by Tarahumara \ optimum germination occurs at 17^0C at night and 20^0C daytime maximum.

Mammillaria magnimamma
Mammillaria centricirrha
Mammillaria gladiata
Neomammillaria magnimamma

Mexico \ small, clump-forming cactus to 300mm tall and spreading to 600mm \ semi-hardy to frost \ cream, pink or red flowers to 20mm wide develop into an edible fruit.

Mammillaria mammillaris
Mammillaria simplex

Venezuela, Tobago, Trinidad, Lesser Antilles, Curacao \ occurs naturally at medium elevation \ small, solitary, cylindrical cactus to 60mm wide 200mm tall \ does not tolerate frost and is best kept above 5^{0}C \ edible fruit \ stem yields a milky juice which is claimed to be edible \ optimum germination between 20^{0}C at night and 30^{0}C daytime maximum.

Mammillaria meiacantha
viejito
chilies on a plate
South west USA, Mexico \ small cactus \ very hardy, often snow covered for weeks and able to withstand temperatures

as low as minus 23^{0}C \ oblong, scarlet, edible fruit.

Mammillaria parkinsonii
Mammillaria morganiana
owl's eye

Mexico \ small, clustering, woolly cactus with stems about 100mm wide \ needs alkaline soil and full sun \ does not tolerate frost and is best kept above 5^{0}C \ edible fruit \ optimum germination between 20^{0}C at night and 30^{0}C daytime maximum.

Mammillaria senilis
Mamillopsis senilis

whichuriki
hikuli rosapara
hikuri
peyote de San Pedro
cabeza de veijo

Mexico \ small, round or slightly elongated, woolly, clumping cactus \ each stem grows up to 150mm long and 55mm wide \ needs full sun \ needs to be kept very dry in winter \ very hardy to frost down to minus 12^{0}C \ used as false peyote by Tarahumara \ optimum germination between 20^{0}C at night and 30^{0}C daytime maximum.

Mammillaria tetrancistra
Phellosperma tetrancistra
fish hook cactus
California \ range from 50-250mm tall and up to 100mm diameter \ sometimes clustering \ needs full sun \ depending on provenance is tolerant of occasional low frost above minus 4^{0}C or hardy to frequent, moderate frost to minus 8^{0}C \ prized as a food by Californian tribes \ whole plant dipped in boiling water then skinned, chopped and added to stew \ the rain god was believed to cut off the rain if the people ate too many plants (conservation through legend) \ optimum germination between 20^{0}C at night and 30^{0}C daytime maximum \ remove from closed, humid environment a few days after germination.

Mammillaria vivipara
Southern North America \ small cactus \ edible fruit.

Mammillaria wrightii
Neomammillaria wrightii
thorny cactus

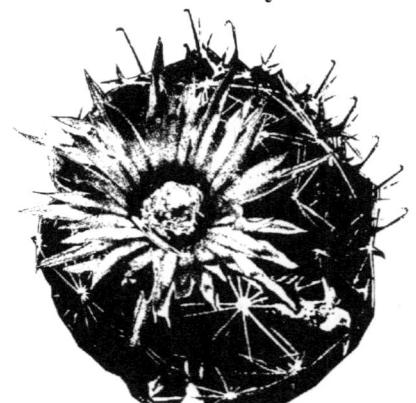

USA, Mexico \ small, columnar, solitary (occasionally offsetting) cactus to 75mm diameter \ full sun \ very hardy to frost down to minus 12^{0}C with some provenances being snow covered for weeks and able to withstand temperatures as low as minus 23^0C \ day flowering during summer \ stem and ripe fruit eaten by Ramah Navaho tribe \ grows from seed at 20-30^{0}C with 13-14 hours daylight per day.

Matucana madisoniorum

Matucana pujupatii
Peru \ probably contains mescaline.

Melocactus spp.

Mexico, Central America, Caribbean, western and northern South America \ 40 species in genus.

Melocactus curvispinus

Melocactus dawsonii
Melocactus delessertianus
Melocactus jakusii
Melocactus loboguerreroi
Melocactus maxonii
Melocactus obtusipetalus
Melocactus ruestii
Cactus caesius
              barba de viejo
Mexico, Central America, Cuba, Trinidad, Tobago, Colombia, Venezuela \ sweet fruit is edible.

Melocactus intortus

Melocactus coronatus
Melocactus ruestii
              melon cactus
          Turk's cap cactus
Central America, Caribbean \ rounded, barrel-shaped cactus to 1m tall \ frost tender \ pink flowers preceed small, edible fruit \ juice is potable.

Mila spp.

Peru \ 3 species in genus.

Mila caespitosa

Mila albisaetacens
Mila cereoides
Mila fortalezensis
Mila pugionifera
Mila sublanata

Central Peru (endangered) \ low (150mm), semi-prostrate, cactus, to 30mm in diameter, growing in clumps of 50 or more stems \ high elevation \ full sun \ day flowering in summer \ tiny, greenish-brown, berry-like fruit is edible.

Myrtillocactus spp.

Mexico, Guatemala \ 4 species in genus.

Myrtillocactus cochal

Cereus cochal

Mexico \ columnar, freely-branching, shrubby cactus with a short trunk about 150mm wide \ grows to 3m tall \ full sun \ day flowering in early summer \ small, purplish-black,

berry-like fruit resembles black currants and is eaten fresh or dried.

Myrtillocactus eichlamii

Guatemala \ columnar, freely-branching, shrubby cactus \ small, edible, purplish-black fruit resembles black currants.

Myrtillocactus geometrizans

Cereus geometrizans

              garambullo
                pitaya
            blue candle

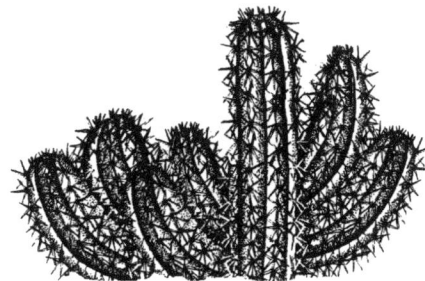

Central Mexico \ columnar, freely-branching, ribbed, tree-like cactus to 4m tall and 3m spread \ medium elevation \ day flowers in early summer \ cultivated for it's small, edible, purplish-blue, olive-shaped berries with a flavour that resembles blueberries \ dried fruit looks and is used like raisins \ commonly planted as a hedge.

Myrtillocactus schenckii
Cereus schenckii

Mexico \ columnar, freely-branching, tree cactus to 5m tall \ full sun \ day flowering in summer \ small, edible, purplish-black fruit resembles black currants.

Neobuxbaumia spp.
Mexico \ 8 species in genus.

Neobuxbaumia tetetzo
Neobuxbaumia tetazo
Pilocereus tetetzo

tetecho

Mexico \ tall, erect, columnar, tree cactus to 15m tall and 300mm in diameter often forming large impenetrable clumps\ seed ground and used as a cereal \ stems chopped as emergency stock feed.

Neoraimondia spp.
Bolivia, Peru / 2 species in genus.

Neoraimondia arequipensis
Neoraimondia aticensis
Neoraimondia gigantea
Neoraimondia macrostibas
Neoraimondia peruviana
Neoraimondia roseiflora
Peru \ used as an ingredient in the inebriating beverage *cimora*

Neoraimondia herzogiana
Neocardenasia herzogiana

strawberry pineapple fruit
Peru \ large branching shrub to short-trunked tree \ fruit claimed to be the best flavoured of the cactus fruit, tasting like strawberry and pineapple.

Neowerdermannia spp.
Peru, Chile, Argentina \ 3 species in genus.

Neowerdermannia vorwerkii
Weingartia vorwerkii

Bolivia, Argentina \ low globular cactus to about 80mm in diameter \ deep taproot \ sunny site \ boiled plant eaten like potatoes.

Obregonia denegrii
**pine cone cactus
obregona
obregonita**

Tamaulipas in Mexico where it is considered vulnerable \ only species in genus \ Mexican Indians held plant as sacred as it contains alkaloids which induce visions.

Opuntia spp.
prickly pear
Probably one of the most used genus of the CACTACEAE family, encompassing 362 species, throughout the Americas. There is certainly much confusion about their nomenclature and hybridisation often occurs between species. The natural distribution of this genus is from British Colombia in Canada to the southern tip of· South America. Some species are hardy to minus 20^{0}C. They range from large, evergreen, tropical trees to small, alpine ground cover. They are generally credited with increasing soil humus content over time.
The fruit of many species known as tunas was eaten by different Indian tribes. Even today the fruit is a staple to many people. Navajo knocked fruit down with a forked stick, whereas Apaches used wooden tongs. Pawnees cooked dried fruit with meat or made it into

a sauce by boiling it in water for 10-12 hours and allowing it to ferment for a while.

The cultivation of some species of *Opuntia* have spread around the world. In Sicily, the practice called *scozzolatura* has been developed. The first crop of flowers are knocked off. The plants flower again in about a month and the following crop although fewer are much larger. Logically, doing this to half of the plants would extend crop period. Fruit pulp with seed removed can be evaporated into *queso de tuna* which resembles cheese and keeps for a long time. *Miel de tuna* is made from crushed fruit pulp, after the seeds are removed. It is boiled slowly into a thick syrup which eventually crystallises to form a sugar similar in taste to maple sugar but with a finer grain.

The seed from the fruit were lightly toasted and ground into a flour-like meal for cakes and bread. Dried seed could be stored for times of scarcity. The pads were roasted in the hot ashes enabling easy removal of skin and thorns. The resulting sweet mucilage was eaten and relied on as an emergency food. The pads can be eaten cooked in batter. The mucilage in the pads has many applications in cooking. It is also used in building, being mixed with whitewash to act as a bonding agent. The plant can be composted to make a fertiliser. Many of the O*puntia* have medicinal uses. The Blackfoot tribe lacerated warts and moles then applied the fuzz from the fruit to the wound as a means of removal. Cactus juice from the pads was boiled in water for pulmonary complaints. The baked joints were opened and applied to fresh wounds.

In Africa a pad is placed on the ground outside the hut window at night as a burglar alarm \ Propagation of *Opuntia* is simple. Break off a pad, allow a day or two to callus, then place it in the ground. Thornless opuntia is used extensively as an animal fodder. It can be established and grown through good seasons and harvested as feed during bad seasons. In Tunisia 100 000 hectares of thornless *Opuntia* plantation saved about 230 000 sheep and goats from starvation. The value is estimated at $8million. The value in reduced suffering cannot be quantified.

A major pest of this genus is the woolly aphid known as cochineal or cochinilla *(Dactylopius opuntia, syn. Dactylopius cochenillifera)*. This insect is the source of raw material of the pigment carmine which was used in pre-Columbian times by the Aztecs and others and has been the centre of a major industry. Cochineal, the dried body of the insect is valued as a red dye. The dye has uses also in food colouring. Modern synthetic dyes have frozen this industry but this is probably temporary because modern toxic chemicals are being replaced by natural equivalents. The mature insect is about the size of a grain of rice. They are carefully brushed from the cactus into bags two or three times a year depending on climate. They are killed by immersion in boiling water then dried in the sun or over a fire. In a dried state they will keep in good condition for many years.

Two types of insect occur. Silver cochineal is greyish-red with a white down covering and is considered more valuable than black cochineal, which has a dark red-brown body and no down. The female is wingless and the male has wings.

silver cochineal on opuntia plant

Generally the best environment for cochineal to thrive is where branches are close together to offer protection. Branches hanging onto the ground allow easy passage of ants which tend to remove predators of the cochineal.

The other serious pest to the *Opuntias* is the prickly pear moth *(Cactoblastis cactorium)*. Egg sticks (10-25mm long), resembling a blade of grass protrude from the cactus pad at right angles. Hundreds of larvae emerge from a single stick. The larval stage of the moth is the source of destruction. It can kill a plant within a few months. The larvae burrow into the plant and cause the leaf to collapse.

They proceed to work through the entire plant.

This insect has been introduced to many countries to control ferrel populations of O*puntias*. These O*puntias* were generally introduced as host to the cochineal insect, but either became very rampant or were abandoned as synthetic dyes became popular and cochineal was no longer economic to harvest. The natural stain carmine (carminic acid) is derived from cochineal. Health problems associated with synthetic colourings are promising to create a new demand for cochineal.

Where prickly pear moth is a risk it is necessary to look for egg sticks which should be removed and destroyed. Two generations occur in a year in most climates (spring and late summer).

Plants infected with larvae should be pruned back to unaffected parts and the infected matter destroyed - possibly a role for pigs.

Optimum germination of seed is between 17^0C at night and 38^0C daytime maximum at 13-14 hours daylight per day. Some species are slow to germinate and are assisted by wet then dry stratification. Soaking in warm distilled water for 12 hours before planting can also be effective.

Opuntia acanthocarpa
Cylindropuntia acanthocarpa

buckhorn cholla

USA to Mexico \ erect to spreading low shrub or small tree, usually growing to 1-2m but occasionally attaining 4m \ it can form extensive forests in some desert conditions \ altitude range from 300-1300m \ very hardy to frost down to minus 12^0C fruit eaten fresh or dried \ stems an important and reliable food source to Cahuilla \ calyxes were pit roasted with inkweed and eaten fresh or dried by Pima \ stem ash to treat burns and cuts \ plant used for stomach ailments \ has been found to contain 0.01% mescaline.

Opuntia amyclaea
Mexico \ cultivated in Mexico and the Mediterranean region for it's fleshy, yellow, edible fruit.

Opuntia arborescens
entraña
jo
chandelier cactus
cane cactus

New Mexico \ fruit picked with tongs and rubbed with a stone to dislodge the spines then boiled and eaten, usually with porridge.

Opuntia arbuscula
Cylindropuntia arbuscula
Cylindropuntia vivipara

pencil cholla

USA, Mexico \ fruit eaten by Seri.

Opuntia arechavaletae
Brazil, Uruguay, Paraguay \ edible fruit.

Opuntia auberi
Nopalea auberi

Mexico, Cuba (rare) \ young pads eaten as a vegetable.

Opuntia austrina
Opuntia pollardii
Opuntia polycarpa

southern prickly pear

Florida \ erect or spreading to 1m \ plant dies back after a couple of years and then re-shoots from the base \ very hardy, often snow covered for weeks and able to withstand temperatures as low as minus 23^0C \ ripe fruit eaten raw \ mucilaginous joints are eaten raw or cooked \ inside of split joint used as a poultice on wounds and cuts.

Opuntia azurea

Mexico \ oval shaped bluish-green joints to 150mm long on a short stem \ full sun \ tender joints eaten as a vegetable \ day flowering in summer \ edible fruit.

Opuntia basilaris
Opuntia whitneyana
beavertail prickly pear

South western USA, Mexico \
low-growing (to 1m) cactus
with flat, oval shaped pads to
200mm long with fine short
spines \ reaching a spread of
1m \ medium to high elevation
\ full sun \ very hardy to frost
down to minus 12^0C \ day
flowering in early summer \ a
major cactus of south west
Californian Indians \ fruit
cooked unripe to a sauce
consistency \ ripe fruit eaten
raw, dried and made into
preserves \ Indians broke
young fruit off with a stick then
cleaned the fine short spines off
with grass or a bunch of sticks
\ flowers and flower buds eaten
cooked \ young pads sliced,
cooked and eaten as substitute
for beans \ young pads were
sun dried after spines were
brushed off \ dried pads keep
indefinitely and can be eaten
after boiling in salt \ seed eaten
in broth \ pulp of older pads
scraped out and used as a wet
dressing (for cuts and wounds)
which needs frequent changing
but helps in healing and
deadens pain \ used to treat
warts \ has been found to
contain 0.01% mescaline.

Opuntia bigelovii
Cylindropuntia bigelowii
jumping cholla
teddy bear cholla

Texas, New Mexico, Central
Mexico \ erect trunk to 1m
high with numerous short
spreading branches \ full sun \
withstands temperatures down
to minus 24^0C for many years
without damage \ buds eaten
cooked or dried for indefinite
storage \ berries and stems
once an important and
dependable food source of the
Cahuilla \ plant used as a
diuretic \ stock eat the leaves,
despite the thorns, once
familiar with the plant \ fruit
contains few if any seed and
most reproduction in the wild
is from fallen joints.

Opuntia boldinghii
Southern Caribbean, Venezuela
\ edible fruit.

Opuntia bonplandii
Ecuador \ edible fruit.

Opuntia brasiliensis
Opuntia bahiensis
Brasiliopuntia brasiliensis
Brasiliopuntia subacarpa
Argentina, Bolivia, Peru,
Paraguay, Brazil \ tree-like \
vigorous grower \ requires
warm conditions \ edible fruit.

Opuntia camanchica
Maihueniopsis camanchica
round leaved cactus
prickly pear cactus
New Mexico \ very hardy,
often snow covered for weeks
and able to withstand
temperatures as low as minus
23^0C \ edible fruit.

Opuntia caracassana
Opuntia wentiana
Southern Caribbean, Venezuela
\ edible fruit.

Opuntia caribaea
Cylindropuntia caribaea
Caribbean, Venezuela \ edible
fruit.

Opuntia chamacuera
Nopalea chamacuera
Edible fruit \ edible pads.

Opuntia chata
Mexico \ edible fruit.

Opuntia chavena
Mexico \ eaten as a vegetable.

Opuntia chichensis
Opuntia ferocior
Cumulopuntia boliviana
Tephrocactus chichensis
Tephrocactus ferocior
Bolivia to Argentina \ low
cactus \ edible fruit.

Opuntia chlorotica
Opuntia palmeri

South west USA, Mexico \ tree-like to 1m tall with joints to 200mm long \ full sun \ very hardy to frost down to minus 12^0C \ day flowering in summer \ edible fruit.

Opuntia clavata
Corynopuntia clavata

South west North America \ low-growing, clumping, mat-forming cactus \ rarely growing taller than two or three joints \ this is a very slow-growing plant which takes root as it spreads sideways \ best in full sun \ slightly calcareous soil \ very hardy, often snow covered for weeks and able to withstand temperatures as low as minus 23^0C \ yellow flowers produce an edible fruit.

Opuntia clavellina
Cylindropuntia clavellina
Mexico \ fruit eaten roasted.

Opuntia cochenillifera
Nopalea cochenillifera
Mexico, Central America \ branching, shrub cactus to 2-4m tall having joints to 250mm long and 100mm wide \ best in full sun \ day flowering in summer \ edible fruit \ young joints pickled \ used extensively as host to the cochineal insect, the source of the red cochineal dye \ used as cattle fodder.

Opuntia crassa
Mexico \ cultivated for edible fruits.

Opuntia cymochila
Opuntia mackensenii
USA \ needs alkaline soil \ fruit used in jelly.

Opuntia cylindrica
Austrocylindropuntia cylindrica
Austrocylindropuntia intermedia
Cylindropuntia intermedia
Peru, Ecuador \ stems can contain 0.9% by dry weight mescaline, Smith claims this is not true.

Opuntia crystalenia
Mexico \ edible fruit.

Opuntia dejecta
Nopalea dejecta

Cuba \ common in cultivation in tropical America \ well armed shrub cactus with

slightly drooping branches \ edible fruit \ stripped joints eaten as a vegetable.

Opuntia delaetiana
Paraguay, Argentina \ weedy plant \ edible fruit.

Opuntia echinocarpa
silver cholla
USA, Mexico \ multi-branched bush \ very hardy to frost down to minus 12^0C \ flowers eaten cooked or roasted.

Opuntia elata
Paraguay, Cuba \ tall bushy cactus \ 250mm long pads to 150mm wide \ best in full sun \ day flowering in late summer \ edible fruit.

Opuntia elatior

Costa Rica, Panama, Colombia, Venezuela, Caribbean \ well armed \ cultivated for its dull-purple, edible fruit \ plant used as an antiseptic, expectorant and to treat biliousness, boils, coughs, guinea worms, inflammation, ophthalmia, pertussis, spasm and sores \ cultivars exist with minimal spines.

Opuntia engelmannii

South west USA, Mexico \ bushy cactus to about 2m tall \ pads to 300mm long \ full sun \ day flowering in early summer \ joints eaten fried \ fruit eaten raw or cooked and good in jelly \ pads as stock feed.

Opuntia ficus-indica

prickly pear
Indian fig
tuna

USA, Mexico \ to 5m tall with a spread of 3m \ introduced to African coast to provide sailors with fruit to prevent them contracting scurvy on their long voyages to India \ today widely cultivated throughout the world for fruit production \ generally spineless varieties are desired in cultivation \ full sun \ well-drained soil \ bright yellow, funnel-shaped flowers produce fruit ranging in colour from white, yellow-white, pink-white, green, brown-green through to blood red depending on the variety \ widely cultivated for edible fruit which is eaten after carefully removing spines and peeling \ syrup made by boiling peeled fruit then straining out the seed (which is useful as a stockfeed) \ a dark red to black paste called *quesco de tuna* was made by further drying the syrup \ jam is made from the fruit with or without peel \ syrup from the red fruit is preferred as a topping to ice cream \ red fruit also preferred in winemaking \ wine can be distilled to make *mampoer* \ young pads, before the spines have hardened, (*nopalitos*, *nopales*, or *pencas*) are sliced and crystallised in sugar or boiled and eaten like snap beans and also pickled as gherkin substitutes (as are immature fruit at 10-12mm) \ young leaves can also be baked with other vegetables in a roast or eaten raw in salads \ in South Africa the peeled fruit is boiled and fermented with sugar and yeast to make a beer \ although low in protein the leaves are useful as a stockfeed particularly where pruning wastes are available or as an emergency fodder \ seed was stored and ground as flour to make *atole* (a thick broth) \ mature fleshy pads were split, soaked and used as a poultice for bruises \ plant used to treat radiation burns, sunburn, scalds, leprosy, tumours, corns and calluses, measles, piles, diarrhoea, diabetes, kidney ailments, wounds and sores, and also as an emollient, decongestant, internulcer and favus \ popular hedge plant \ leaves mixed with caustic soda, water and fat then cooked to make a soap \ pads crushed and boiled yield a sticky juice which was added to whitewash and mortar to improve adhesiveness \ cultivars exist.

Opuntia floccosa
Opuntia atroviridis
Opuntia crispicrinitus
Opuntia cylindrolanata
Opuntia pseudo-udonis
Opuntia rauhii
Opuntia tephrocactoides
Opuntia udonis
Opuntia verticosa
Austrocylindropuntia floccosa
Austrocylindropuntia lauliacoana
Austrocylindropuntia machacana
Austrocylindropuntia tephrocactoides
Tephrocactus atroviridis
Tephrocactus crispicrinitus
Tephrocactus floccosus
Tephrocactus pseudo-udonis
Tephrocactus rauhii
Tephrocactus udonis
Tephrocactus verticosus

Central Bolivia, Peru \ low cactus to 300mm tall and 80mm thick, forming dense cushions to 2m wide \ altitudes 4,000-4,500m \ full sun on slightly calcareous soil \ day flowering in summer \ edible fruit \ flowers eaten cooked or roasted.

Opuntia fragilis
thimble
brittle prickly pear

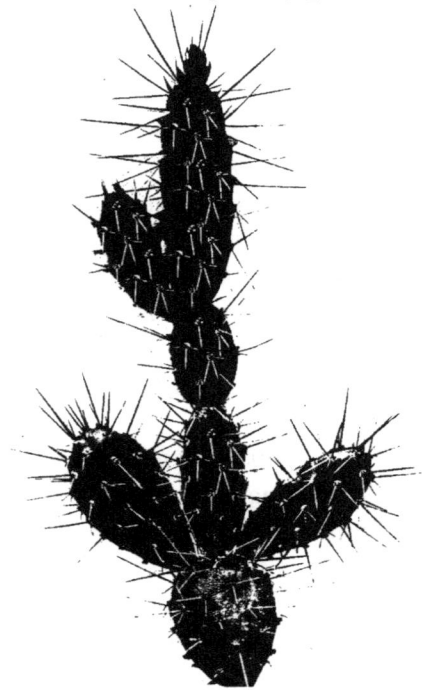

Western Canada, USA \ sandy and gravelly soils \ altitudes to 2400m \ withstands temperatures as low as minus 40^{0}C in western Canada \ edible fruit \ stem flesh boiled with fat into a soup \ flesh eaten roasted or pit-cooked \ flesh eaten to stop urination \ heated quills applied as a poultice to cuts, sores, boils or swollen throats \ mucilaginous material from cut stems used as glue.

Opuntia fulgida
Cylindropuntia fulgida
jumping cholla
chain-fruit cholla
boxing gloves

South west USA, Mexico \ prolific across the mainland region of the Sonoran Desert \ tree to 4.5m tall \ some types are almost thornless (eg. *var. mammillata*) whereas others are well armed \ full sun \ slightly calcareous soil \ day flowering in mid summer \ chains of green, fleshy, edible, juicy fruit \ fruit lacks spines and grows to 25-40mm long \ hardy to frequent, moderate frost to minus 8^{0}C \ fruit is extremely variable in size and flavour and the Seri tribe cultivated select plants with quality fruit \ pendulous chains of fruit persist on the plant for years \ gummy sap which exudes from cut stems of plant is eaten raw or roasted and also used as a beverage when mixed with water \ plant used to treat toothache, diarrhoea and short-windedness \ gum from stems used as a size and fabric stiffener \ pads fed to livestock \ wood for small items.

Opuntia fuliginosa
Mexico \ edible fruit.

Opuntia humifusa
Opuntia allairei
Opuntia calcicola
Opuntia compressa
Opuntia cumulicola
Opuntia fuscoatra
Opuntia humifusa
Opuntia impedata
Opuntia nemoralis
Opuntia rafinesquil
Opuntia rubiflora
Opuntia vulgaris
devil's tongue
western prickly pear
eastern prickly pear
Joseph's coat

Southern USA, Mexico, Cuba \ forms clumps \ joints to 160mm long and 100mm wide \ best in full sun \ hardy \ naturally occurs on sterile sand of the Atlantic seaboard \ day flowering in full sun \ cultivated for edible fruit which is eaten raw, stewed or dried for winter \ young joints impart a mucilaginous quality to soups and stews.

Opuntia hyptiacantha
Opuntia cretochaeta
Opuntia matudae
Mexico \ edible fruit with a sweetish, acidic, purple flesh is eaten raw or cooked \ young joints cooked as vegetable.

Opuntia hystricina

na'vu

USA, Mexico \ low-growing, clump-forming cactus \ joints are almost circular, about 120mm long and up to 100mm across \ full sun \ day flowering in summer \ fruit traded to Hopi from Havasupi.

Opuntia imbricata
Opuntia arborescens
Cylindropuntia imbricata

coyonostole
cholla
cane cactus
walking stick cactus
DeCandolle tree cholla
devil's rope pear
rose pear

Mexico and southern USA \ small tree or thicket-forming shrub to 3m tall \ mostly in grassland but also in woodland (especially mesquite *(Prosopis spp.)* thickets) and chaparral \ gravelly or sandy soil on hills, plains and valleys \ 600-2400m altitude \ very hardy, often snow covered for weeks and able to withstand temperatures as low as minus 23^0C \ best in full sun \ red-purple flowers occur during the day in summer \ yellow, edible fruit up to 50mm diameter \ unopened flower buds are steamed or boiled then added to stews and salads \ they are less mucilaginous when dried \ dried fruit boiled with leaves of *Atriplex elegans, Atriplex coranata or Atriplex serenana* to counteract its acid flavour \ Pima pit-baked fruit overnight then dried it as store for winter \ Lagona, Acoma and Keres ate the roasted joints \ joints were split lengthwise, sun-dried and stored as food \ ground needle coverings were made into a paste and applied to boils \ stem pith used dried for running ear and ear ache \ reported to contain Mescaline \ thorn as tattooing and sewing needle \ dried woody stems used for torches and candles \ flowers were an indicator to Keres of time to plant beans \ planted as fences by some tribes as spines are especially painful.

Opuntia inamoena
Opuntia quipa
Platyopuntia inamoena

Brazil, low-growing to 600mm \ pads to 160mm long, 75mm wide and 20mm thick \ best in full sun \ day flowering in summer \ edible fruit.

Opuntia joconostle
Mexico \ cultivated for fruit \ fruit pulp and especially the peel used in candy making.

Opuntia karwinskiana
Nopalea karwinskiana
Mexico \ shrub cactus \ young joints eaten as a vegetable.

Opuntia laevis
South western USA \ very hardy to frost down to minus 12^0C \ valued as edible fruit \ stems eaten diced (after spines are removed) in salad with a vinegar marinade.

Opuntia lanceolata
Origin obscured by wide cultivation for edible fruit \ tree-cactus to 2m tall \ joints to 300mm long and 80mm wide \ full sun \ day flowering in early summer.

Opuntia leptocaulis
Cylindropuntia brittonii
Cylindropuntia leptocaulis
desert Christmas cactus
pencil cholla
tasajillo
tesajo

Southern USA into Mexico \ bushes or erect shrubs (0.5-7m tall) \ heavy, bottom-land soils \ 60-1500m altitude \ hardy to frequent, moderate frost to minus 8^0C \ one form has short spines \ full sun \ day flowering in mid summer \ Pima ate raw fruit \ fruit is bright red or sometimes yellow and persists on the plant throughout winter \ yellow fruit have been observed to be preferred by birds in the wild \ fruit claimed to be slightly hallucinogenic.

Opuntia leucotricha
duraznillo
Mexico \ freely-branching, tree-like cactus to 4m tall \ full sun \ day flowering in mid summer \ white or red fruit eaten raw or made into a beverage called *colonche* which has been made for more than 2000 years.

Opuntia lindheimeri
Opuntia aciculata
Opuntia tardospina

South western USA, Mexico \ shrubby cactus to about 2m tall with pads up to 200mm long \ full sun \ very hardy to frost down to minus 12^0C \ day flowering in early summer edible fruit \ tender joints as a vegetable \ plant used to treat dyspepsia, mumps, swelling and in veterinary use to treat bruises \ joints fed to cattle after burning off spines.

Opuntia linguiformis
lengua de vaca
cow's tongue
South western USA \ upright to sprawling cactus with joints to 400mm long and 100mm wide \ full sun \ day flowering in early summer \ fruit used to make jam and jellies.

Opuntia macrocentra
Opuntia gosseliniana
Opuntia violacea
black spined prickly pear
purple prickly pear
blue blade
dollar cactus

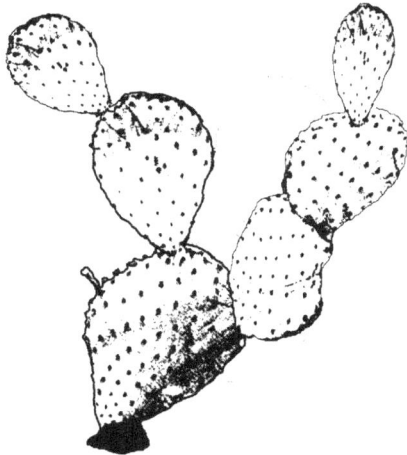

South western USA, Mexico \
low-growing, spiny cactus to
1m tall and spreading to 3m \
flat pads grow to about 150mm
long and 10mm thick \ medium
elevation \ needs full sun \ very
hardy, many provenances often
snow covered for weeks and
able to withstand temperatures
as low as minus 23^0C \
spineless, purple, edible fruit
grows to 60mm long \ good in
a medium height hedge.

Opuntia maxima
Origin obscured by widespread
cultivation for it's edible fruit.

Opuntia megacantha
nopal
mission tuna
mission prickly pear

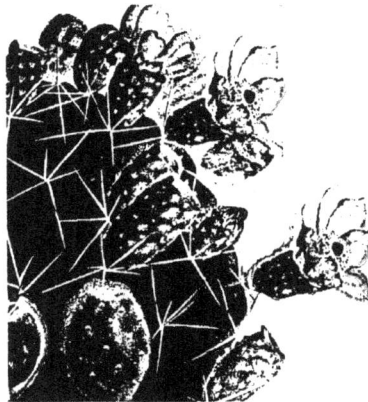

Mexico \ cultivated widely for
edible, green-coloured fruit \
fruit consumed raw, cooked, in
sweetmeats and syrups \
nochote is the fermented fruit
juice mixed with pulque and
water \ dried fruit made into a
flour to make cakes \ tender
joints eaten cooked as a
vegetable \ plant used to treat
inflammation and used as a
laxative and poultice \ stem
juice mixed with tallow to
make candles.

Opuntia megarhiza
Mexico \ plant used to treat
fractures and inflammation.

Opuntia monacantha
Platyopuntia
brunneogemmia
Platyopuntia vulgaris
drooping tree pear
smooth-leaved tree pear
Brazil, Uruguay, Paraguay,
Argentina \ erect, tree cactus
can exceed 2m tall \ long spines
which diminish in number on
the trunks of older plants \ in
Australia they became dense
stands to 10m tall in the
brigalow *(Acacia harpophylla)*
forming impenetrable barriers \
pear-shaped to spherical fruit is

red with a rich red fruit
containing minimal seed \
commonly cultivated as host to
cochineal insect.

Opuntia moniliformis
Opuntia haitiensis
Opuntia picardae
Opuntia testudinis-crus
Consolea moniliformis

Haiti, Dominican Republic,
Puerto Rica \ tree cactus \
edible fruit \ plant used to treat
tumours.

Opuntia occidentalis
Mexico, south western USA \
considered a hybrid of other
Opuntia spp. \ valued for its
edible fruit which should be
peeled carefully after spines are
removed \ buds eaten in soups \
body grated as a garnish and
used in pickles.

Opuntia pachona
Mexico \ eaten as a vegetable.

Opuntia paraguayensis
Opuntia bonaerensis
Opuntia chakensis

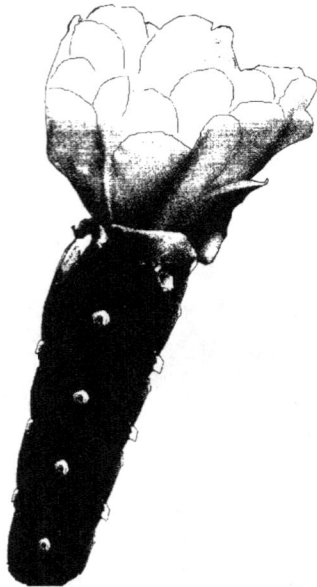

Paraguay, Uruguay, Northern Argentina \ bushy, semi-erect cactus with oblong joints growing to 200mm long and 150mm wide \ full sun \ day flowering in mid summer \ fruit eaten raw and in salads / used as a living fence.

Opuntia parryi
Cylindropuntia parryi

cane cholla

USA, Mexico \ green, tan or brown berries eaten \ stems are eaten.

Opuntia phaeacantha
Opuntia angustata
Opuntia woodsii

purple fruited prickly pear bastard fig

South western USA, Mexico \ low-growing, sprawling or erect, freely-branching, spiny cactus to 1.5m tall with a spread of 2.5m \ medium to high elevations (up to 2500m) \ very hardy, many provenances often snow covered for weeks and able to withstand temperatures as low as minus 23^{0}C \ edible fruit grow to 60mm long \ joints (up to 150x100mm) are eaten roasted as a vegetable or baked with sugar, cinnamon and butter as a desert, seed are dried, parched and ground into a flour for use in cakes and gruel \ pads used as stock fodder.

Opuntia pilifera

Mexico \ tree cactus to 4m tall \ oblong joints can reach 300mm long and 100mm across \ full sun \ day flowering in summer \ edible fruit.

Opuntia plumbea

wide cactus prickly pear

Mexico \ fruit picked with a split stick and spines removed by singeing in the fire or with juniper boughs, then eaten raw, dried or boiled \ fleshy joints singed in fire for spine removal then cooked on coals \ mucilaginous material from peel, roasted stem was used to lubricate the hand of a midwife removing a placenta \ cactus spines used for lancing small skin abscesses and piercing ears.

Opuntia polyacantha
Opuntia arenaria
Opuntia nicholii

yz:'ngu many spined opuntia

Alberta, Canada to Southern USA \ very hardy, many provenances often snow covered for weeks and able to withstand temperatures as low as minus 23^{0}C \ fruit picked with wooden tongs and rolled in the sand until the spines are removed then eaten raw \ fruit also dried and cooked later to make stews and soups \ fruit traded to Hopi from Havasupi \ young joints eaten boiled or fried \ joints have larger thorns removed by burning and are then boiled and the remaining thorns washed off \ joints are then dipped into a syrup made from sweet corn \ infusion of stems for diarrhoea \ plant used also to treat backache, moles, warts and wounds.

Opuntia pottsii
Opuntia ballii
Opuntia delicata
Opuntia setispina
Opuntia tenuispina

prickly pear

South western USA, Mexico \ very hardy, many provenances often snow covered for weeks and able to withstand temperatures as low as minus 23^0C \ stems boiled, juice squeezed out and used as a binding agent (like egg white) to make tortillas \ fruit is smaller and juicier than most with a rich, reddish-yellow colour \ juice fermented to make a superior cactus wine.

Opuntia pseudo-tuna
Plant used to treat tumours.

Opuntia rastrera

Mexico \ edible fruit \ young pads eaten as a vegetable.

Opuntia reflexispina
Corynopuntia reflexispina
Mexico \ plant used to treat diarrhoea.

Opuntia robusta
Opuntia guerrana

dinner plate
wheel cactus

Mexico \ strong-growing, tree-like cactus to 5m tall \ pads are round or even broader than long and up to 450mm in diameter \ full sun \ yellow flowers appear in the day in summer \ red fruit eaten raw, cooked or made into sherbet \ young joints eaten boiled as a vegetable and diced in coleslaw used as a living fence.

Opuntia soederstromiana
Opuntia dobbieana
Ecuador \ edible fruit.

Opuntia soehrensii
Opuntia boliviensis
Opuntia cedergreniana
Venezuela, Bolivia, Trinidad \ fruit used as a food colouring \ grown as a hedge plant.

Opuntia streptacantha

tuna cardona
cardona
white-spined pear

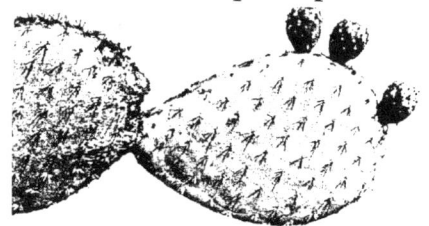

Mexico \ cultivated for it's plum red or dark purple fruit which is eaten raw, cooked, in preserves and used in *colonche* (a fermented drink), wine and *quesco de tuna* (cactus pear cheese) \ young pads used as a vegetable \ good hedge plant.

Opuntia stricta, var. dillenii

Opuntia anahuacensis
Opuntia atrocapensis
Opuntia bahamana
Opuntia dillenii
Opuntia horrida
Opuntia inermis
Opuntia keyensis
Opuntia macrarthra
Opuntia magnifica
Opuntia melanosperma
Opuntia nitens
Opuntia zebrina

**dillen prickly pear
nopal
kartu-patuk
chappal
naphana**

Florida, Caribbean, Surinam (rare), Ecuador, Uruguay \ tall, erect or sprawling, branched cactus to 3m \ pads grow to 250mm long \ thornless types \ full sun \ day flowering in late summer \ ripe fruit eaten raw \ mucilaginous joints are eaten raw, cooked or preserved by sun drying \ one variety is said to have the optimum balance of phosphorous and nitrogen to be ideal for cattle feed \ inside of split joint used as a poultice on wounds and cuts \ juice diluted in water taken as diuretic \ in India where it has become naturalised it is applied to scorbutic ulcers, a hot poultice for extracting Guinea worms and bringing boils and abscesses to a head \ plant used to treat ophthalmia, syphilis, pimples and sores \ became a declared weed in Queensland in 1883.

Opuntia subulata

Opuntia exaltata
Austrocylindropuntia exaltata
Austrocylindropuntia subulata

Southern Peru, Bolivia, Argentina \ branching, tree to 6m tall with trunk about 200mm thick \ cylindrical pads to 100mm long \ full sun \ small round stem tips eaten in salads \ cultivated as a hedge plant.

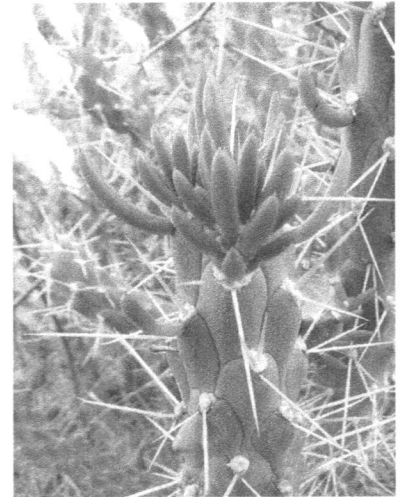

Opuntia subulata

Opuntia sulphurea

Opuntia maculacantha
Opuntia vulpina
Platyopuntia sulphurea

Argentina, Bolivia \ low-growing, straggly cactus \ pads are oval shaped to 200mm long \ full sun \ day flowering in summer \ edible fruit.

Opuntia tomentosa
Opuntia hernandezii
Opuntia macdougaliana
Opuntia sarca
**velvety tree pear
tree pear**

Mexico \ cultivated for edible
fruit which is eaten raw and in
chutneys, jams and jellies \
popular host to cochineal
insect.

Opuntia tuna

**tuna
nopal
panini**

Jamaica, Dominican Republic \
cultivated for edible fruit \ plant
used to treat asthma, diarrhoea,
gonorrhoea, rheumatism and
made into a poultice.

Opuntia undulata
Mexico \ sweet fruit with a
sweet, red pulp.

Opuntia versicolor
Cylindropuntia versicolor
**staghorn cholla
pencil cholla**

South west USA, Mexico \ low
spreading shrub to tree cactus
(1-4.5m tall) \ deep sandy soils
along washes where there is an
abundance of water \ 600-
900m altitude \ full sun \ day
flowering in summer \ edible
fruit is green and usually tinged
with red, purple or lavender
and usually spineless \ fruit
persists in dangling chains on
plant for more than a year \
used as forage for cattle and
deer.

Opuntia whipplei
Cylindropuntia abyssi
Cylindropuntia whipplei
**pi:'nga
cholla cactus**

*above: erect form.
below: erect form
emerging after being
covered by shifting sand
dune.*

USA to Mexico \ erect bush to
2m tall \ joints to 300mm long \
very hardy, often snow covered
for weeks and able to
withstand temperatures as low
as minus 23^0C \ fruit eaten raw
or stewed and eaten with
squash \ fruit traded to Hopi
from Havasupi \ Hopi chewed
the raw root for diarrhoea \
good for sand stabilisation.

Oreocereus spp.
Bolivia, Peru, northern Chile,
north western Argentina \ 11
species in genus.

Oreocereus leucotrichus

*Oreocereus
hendriksenianus
Arequipa leucotricha
Borzicactus
hendriksenianus
Borzicactus leucotrichus*

**chicha-chicha
old man of the Andes**

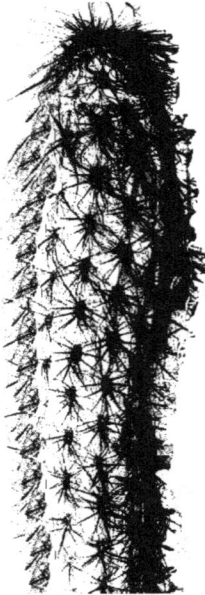

Chile, Bolivia, Peru \ fairly slow growing, large (1m tall, 200mm wide), columnar plant covered in long silky, protective hair \ best in well drained sunny location \ semi-hardy to frost and some provenances can withstand temperatures down to minus 12^{0}C \ day flowering in mid summer \ edible fruit.

Pachycereus spp.

Arizona, Mexico \ 12 species in genus \ propagation from seed at 20-30^{0}C with 13-14 hours daylight per day.

Pachycereus hollianus

*Lemaireocereus hollianus
Stenocereus hollianus*

Mexico \ clumping, branching, tree cactus to 5m tall with branches about 60mm thick \ full sun \ day flowering in early summer \ edible fruit \ extract used as food thickener, adhesive and as a laxative.

Pachycereus marginatus

*Lemaireocereus marginata
Marginatocereus
marginatus
Stenocereus marginatus*

organ pipe cactus

Mexico \ upright sparsely stemmed from 6-15m tall and 300mm thick \ full sun \ day flowering in summer \ edible spiny fruit \ forms an impenetrable hedge.

Pachycereus pecten-aboriginum

*Pachycereus
tehuantepecanus*

**cawe
chawe
wichowaka (insanity)
cardón hecho hecho
Indian comb
bitaya mawalí
chawi-ro-ko
hecho**

Mexico to California \ columnar, tree cactus to 10m with a trunk to 2m tall and 450mm in diameter \ needs full sun \ day flowers occur in summer \ seed eaten slightly browned or made into a flour for cakes \ it is claimed to be a cure for cancer \ Tarahumara express the sap from young branches to make a drink which contains mescaline like San Pedro and another psycho-active compound pectinine \ trunk used in construction.

Pachycereus pringlei
Pachycereus calvus

cardón
cardón pelon
elephant
senita
peyotillo
peyote meco

Mexico \ tree cactus to 20m tall and a 5m spread, with a trunk diameter of 2m \ lives to more than 200 years \ needs full sun \ day flowering in summer \ fruit has slight molasses flavour \ eaten raw or made into a refreshing drink \ seed eaten slightly browned or made into flour \ pulp of ripe fruit mixed with pulp of unripe fruit and kneaded into a mass which is made into cakes and dried \ stem chopped for emergency stock feed \ reliable source of emergency water \ stems used in hut construction.

Pachycereus schottii
Cereus schottii
Lophocereus mieckleyanus
Lophocereus schottii
Lemaireocereus schottii

senita
whisker cactus

South west USA, Mexico \ columnar, rib-stemmed, spineless, tree cactus to 7m tall and spreading to 2m \ does not tolerate frost and is best kept above 5^0C \ flowering tops produce bristles \ nocturnal, funnel-shaped, pink flowers in summer grow to 30mm wide, followed by small, red, edible fruit \ plant crushed and thrown into pools to stun fish.

Pachycereus weberi
Pachycereus gigas
Cereus weberi
Lemairocereus weberi
Stenocereus weberi

cardón
candebobe

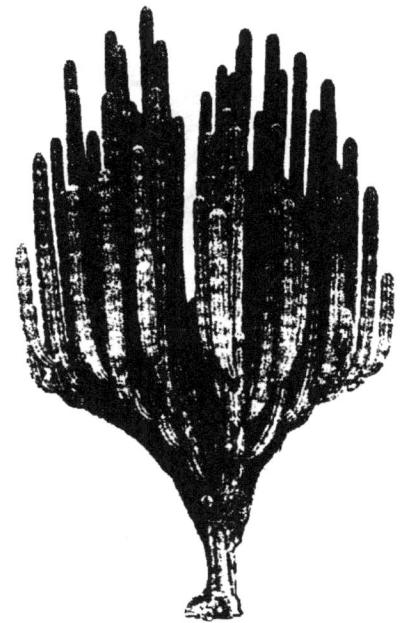

Mexico \ tree cactus to 10m tall with branches about 100mm thick \ edible fruit has spines which fall away when fruit is mature \ seed made into a flour.

Pelecyphora spp.
Northern Mexico \ 2 species in genus.

Pelecyphora asselliformis
peyote
peyotillo
hatchet
peyote meco

Northern Mexico (vulnerable) \ small, grey-green, tufted, cylindriconical up to 100mm in diameter and 100mm tall \ requires full sun \ extracts have antibiotic qualities \ hallucinogenic properties are likely.

Peniocereus spp.
Central America, Mexico and south western USA \ 20 species in genus \ thin stems with large underground tubers \ grows from seed at 20-30^{0}C with 13-14 hours daylight per day \ after germination move into strong light and dry air to avoid rot

Peniocereus greggii
Cereus greggii
Huevo de Venado
deerhorn cactus
night blooming cereus

South west USA, Mexico \ hardy to frequent, moderate frost to minus 8^{0}C \ night flowering \ scarlet, fleshy fruit eaten raw, cooked and made into jam \ large, tuberous root (up to 50kg) boiled, baked, roasted in ashes, fried as chips or parboiled then cooked in fritters \ roots chewed as thirst quencher \ Papago baked the roots whole in the ashes then peeled and ate them \ shoots and stalks eaten as greens \ seed pod mixed with deer grease and used as a salve to be rubbed on sores.

Peniocereus serpentinus
Nyctocereus castellanosii
Nyctocereus serpentinus

Mexico \ slender, cylindrical, clambering cactus \ needs partial shade \ night flowering in summer \ edible fruit.

Peniocereus striatus
Peniocereus diguetii
Cereus striatus
Wilcoxia striata

Mexico, USA \ shrubby cactus with long slender stems to 1m long \ needs shade \ day flowering in summer \ nocturnal flower, edible fruit.

Peniocereus viperinus
Cereus viperina
Cullmannia viperina
Wilcoxia tuberosa
Wilcoxia viperina

Mexico, Texas \ shrubby cactus to 0.3m high with slender, green, erect, snake-like stems small, white spines \ red funnel-like flowers produce an edible fruit.

Pereskia spp.
Widespread from southern USA into South America \ 16 species in genus \ succulent trees, shrubs and climbing plants with flat, succulent leaves become deciduous in cold conditions \ appearance is not typical of cacti \ mostly tropical and frost tender \ require plenty of moisture in hot conditions and minimal moisture in cold \ the leaved species are believed to be the

oldest of the cacti showing how the others have modified to overcome harsh conditions.

Pereskia aculeata
Pereskia pereskia

**Barbados gooseberry
lemon vine
blade apple**

USA to Argentina, including Caribbean (rare in French Guiana, Surinam)\ scrambling, deciduous, fast-growing plant needing support and growing to a height of 10m and width of 5m \ tropical and not hardy below 5^0C \ bright light and some shade \ cultivated \ creamy-white flowers with an orange centre occur in the day in summer \ light-yellow, spherical, edible fruit to 20mm in diameter is sometimes eaten raw but more often stewed or made into jam \ leaves and young shoots in salads, as a potherb and in preserves \

leaves as fodder for livestock \ used in hedges and a cover on walls.

Pereskia bahiensis
Eastern Brazil \ edible fruit.

Pereskia bleo
Cactus bleo
Rhodocactus bleo

spinach cactus

Colombia, Paraguay \ tree-like cactus to 7m tall \ large leaves (to 200mm) are eaten as vegetable.

Pereskia grandifolia
Pereskia grandiflorus
Pereskia tampicana
Rhodocactus grandifolius
Rhodocactus tampicanus

rose cactus

Brazil \ deciduous, erect, bushy cactus (5m tall and 3m wide) with black spines \ full sun and well drained soil \ not hardy to cold below 10^0C \ single, rose-like flowers form in the day in

summer \ edible fruit \ used in hedges \ propagation from cutting or from seed at 20-30^0C with 13-14 hours daylight per day.

Pereskia guamacho
Pereskia colombiana
Rhodocactus colombianus
Colombia, Venezuela, Antilles \ edible fruit.

Pereskia sacharosa
Pereskia moorei
Pereskia saipinensis
Pereskia sparsiflora
Rhodocactus sacharosa
Rhodocactus saipinensis
Argentina, Paraguay, Bolivia, Brazil \ tall, many-branched tree to 8m \ leaves to 100mm long are eaten as a vegetable.

Pereskiopsis spp.
Mexico, Guatemala \ 9 species in genus.

Pereskiopsis aquosa
**tuna de agua
tasajillo**
Mexico \ shrubby cactus \ yellow-green pear shaped fruit used in preserves and drinks.

Pereskiopsis porteri
Mexico \ shrubby cactus \ edible fruit.

Pilosocereus spp.
Mexico, Caribbean, South America \ 64 species in genus \ night flowering.

Pilosocereus chrysacanthus
Cephalocereus chrysacanthus
Cephalophorus chrysacanthus

golden spines
Mexico \ edible fruit.

Pilosocereus leucocephalus

*Pilosocereus
palmeriPilosocereus
maxonii
Cephalocereus
leucocephalus
Cephalocereus maxonii
Cephalocereus palmeri*

Mexico, Guatemala, Honduras \ edible fruit.

Pilosocereus moritzianus

*Pilosocereus backebergii
Pilosocereus claroviridis
Cephalocereus moritzianus*
Venezuela, Trinidad, Tobago \ edible fruit.

Pilosocereus piauhyensis

*Pilosocereus gaturianensis
Pilosocereus mucosiflorus
Cephalocereus piauhyensis
Pseudopilocereus
piauhyensis*
Brazil \ flesh candied.

Pilosocereus robinii

*Pilosocereus bahamensis
Pilosocereus keyensis
 Pilosocereus deeringii
Pilocereus robinii
Cereus robinii var. robinii
Cephalocereus bahamensis
Cephalocereus deeringii
Cephalocereus keyensis*

**key west tree cactus
tree cactus**

Bahamas, Cuba (vulnerable), USA (Florida where it is endangered because of urban development) \ succulent shrub to tree growing to 10m \ grows in dense second growth jungle near the sea and is typically host to Spanish moss *(Tillandsia spp.)* ripe, spineless fruit eaten raw.

Pilosocereus royenii

*Pilosocereus barbadensis
Pilosocereus brooksianus
Pilosocereus millspaughii
Pilosocereus nobilis
Pilosocereus swartzii
Pilosocereus urbanianus
Pilocereus brooksianus
Pilocereus millspaughii
Cephalocereus barbadensis
Cephalocereus brooksianus
Cephalocereus millspaughii
Cephalocereus monoclonos
Cephalocereus nobilis
Cephalocereus royenii
Cephalocereus swartzii
Pseudopilocereus nobilis*

Caribbean \ tall, columnar, cactus to 8m tall and 100mm thick \ freely-branching from the base \ full sun \ day flowering in summer \ edible fruit.

Pilosocereus salvadorensis
Pilosocereus hapalacanthus
Pilosocereus sergipensis
Austrocephalocereus
salvadorensis
Pseudopilocereus
hapalacanthus
Pseudopilocereus
salvadorensis
Pseudopilocereus
sergipensis
Brazil \ edible fruit.

Polaskia spp.
Chichipia spp.
Mexico \ 2 species in genus.

Polaskia chende
Heliabravoa chende
Lemaireocereus chende
Mexico \ tree cactus \ edible fruit.

Polaskia chichipe
Stenocereus chichipe
Lemaireocereus chichipe
chichipe
Mexico \ columnar cactus to 5m tall and 75mm thick \ branching from the base \ full sun \ day flowering in summer \ the red fruit (chichituna) are eaten \ analysis of this species has shown Mescaline, 4-Hydroxy-3,5-dimethoxyphenethylamine and 3,4-Dimethoxyphenethylamine, a compound reported in high concentration in the urine of schizophrenics and lower than normal concentration in the urine of sufferers of Parkinson's Disease \ ash used as a potassium fertiliser.

Rhipsalis spp.
Tropical and sub tropical Americas, Africa, Madagascar, Mascarenes, Sri Lanka \ 40 species in genus.

Rhipsalis baccifera
Rhipsalis bartlettii
Rhipsalis campos-portoana
Rhipsalis cassutha
Rhipsalis cassuthopsis
Rhipsalis cassytha
Rhipsalis cassythoides
Rhipsalis coralloides
Rhipsalis densiareolata
Rhipsalis erythrocarpa
Rhipsalis fasciculata
Rhipsalis heptagona
Rhipsalis horrida
Rhipsalis lindbergiana
Rhipsalis madagascariensis
Rhipsalis minutiflora
Rhipsalis pilosa
Rhipsalis quellebambensis
Rhipsalis shaferi
Rhipsalis undulata
Erythrorhipsalis campos-portoana

mistletoe cactus

Florida, Mexico, Central America, West Indies, South America to Peru, West Africa, Madagascar, Mascarene Islands, Sri Lanka \ pendant or upright epiphyte to 1m long \ thin, cylindrical, segmented stem to 0.3m tall and arching branches with a spread of 1.5m \ light-coloured spines \ moist tropical forests and open sunny sites \ best in partial shade \ drought and frost tender \ day flowering in early summer \ expanding, greenish-white flowers produce an abundance of small, sweet, soft, grape-like, edible fruit which can be white, pink or red.

Schlumbergera spp.
Christmas cactus
Easter cactus

South eastern Brazil \ 6 species in genus \ hybrids of various species are bred as popular indoor plants and are known air filtering plants \ propagation from cutting.

Sclerocactus spp.

Northern Mexico and south western USA \ 23 genus in species.

Sclerocactus uncinatus

Ancistrocactus crassihamatus
Ancistrocactus uncinatus
Ferocactus crassihamatus
Ferocactus mathssonii
Ferocactus uncinatus
Glandulicactus crassihamatus
Glandulicactus mathssonii
Glandulicactus uncinatus
Hamatocactus crassihamatus
Thelocactus crassihamatus
Thelocactus mathssonii
Thelocactus scheerii
Thelocactus uncinatus

South west USA and Mexico (where it is endangered) \ globular to columnar cactus to about 150mm diameter and 200mm tall \ long, curving, hooked spines \ prefers alkaline, gritty soil and sunny site \ hardy to frequent, moderate frost \ edible fruit \ grows from seed at 20-30^{0}C with 13-14 hours daylight per then remove from closed, humid environment a few days after germination into strong light and dry air to avoid rot.

Sclerocactus whipplei

Sclerocactus havasupaiensis
Sclerocactus parviflorus
Ferocactus parviflorus
Ferocactus whipplei
Pediocactus whipplei
Thelocactus whipplei
rose devil claw

South western USA \ short, spiny, cylindrical or globular cactus to 0.25m tall and 0.1 m wide \ very cold hardy some provenances often snow covered for weeks and able to withstand temperatures as low as minus 23^{0}C \ medium elevation to Mediterranean climate \ reddish-green, oblong fruit to 20mm long is edible \ edible seed is released from splits in the base of the fruit.

Selenicereus spp.

Caribbean, tropical America \ genus of 29 species of half-hardy, nocturnal-flowering, climbing cacti.

Selenicereus grandiflorus

Selenicereus hallensis
Selenicereus kunthianus
Cereus grandiflorus
queen of the night
rope cactus
night flowering cactus

Mexico, Belize, Nicaragua, Costa Rica, Caribbean \ variable species is freely-branching, creeping succulent with 7-ribbed stems to 5m long with stems about 20-30mm thick \ Mediterranean to tropical climates \ best in partial shade and well-drained humus rich, slightly acidic soil \ tender to frost \ night flowering in summer \ edible fruit is greenish-red, spiny and 50mm in diameter \ alkaloid similar to digitalin in stems and flowers is used as a heart tonic and to treat angina, dropsy and rheumatism \ used in homeopathic medicines to improve blood circulation \ possible psychoactive effects.

Selenicereus hamatus

Mexico (rare) \ creeping succulent to 4m long with stems to 6mm thick \ semi-shade \ well-drained slightly acidic compost \ white, waterlily-like nocturnal flowers are 200mm across and appear in summer \ edible fruit.

Selenicereus megalanthus
Cereus megalanthus
Mediocactus megalanthus

pitaya amarilla
yellow pitaya

Colombia, Ecuador, Bolivia, Peru \ edible fruit.

Selenicereus pteranthus
Selenicereus nycticalus
Cereus pteranthus
Cereus nycticalus

princess of the night

Mexico \ creeping succulent with four to six angled stem to 40mm thick \ shade \ acid compost \ nocturnal flowering in early summer \ edible fruit.

Selenicereus setaceus
Selenicereus coccineus
Selenicereus rizzinii
Cereus hassleri
Cereus lindbergianus
Cereus lindmanii
Mediocactus coccineus
Mediocactus hassleri
Mediocactus lindmanii
Bolivia, Brazil \ creeping, climbing cactus \ edible fruit.

Selenicereus testudo
Deamia diabolica
Deamia testudo

Mexico, Central America, Colombia \ variable, epiphytic, clambering cactus to 3m long \ partial shade and humid conditions \ porous acid compost \ day flowering in summer \ edible fruit.

Stenocactus spp.
Mexico \ 10 species in genus.

Stenocactus phyllacanthus
Echinofossulocactus phyllacanthus
Echinofossulocactus tricuspidatus
 Ferocactus phyllacanthus

Mexico \ low globular cactus to 80mm diameter \ full sun \ hardy to frequent moderate frost to minus 8^0C \ plant sliced and candied \ optimum seed germination between 17^0C at night and 20^0C daytime maximum at 13-14 hours daylight per day.

Stenocereus spp.
South western USA, Mexico, Caribbean, northern South America \ vary from tree-like thickets 12m or more tall to shrubby thickets of 2 to 3m \ usually night flowering \ does not tolerate frost and is best kept above 5^0C \ grows from seed at 20-30^{0}C with 13-14 hours daylight per day.

Stenocereus alamosensis
Cereus alamosensis
Rathbunia alamosensis
Rathbunia neosonorensis

sina
boa constrictor

North western Mexico \ slender-stemmed, clambering, shrubby, branching cactus to 2m tall and 10m wide \ partial shade \ clambers through tall scrub \ does not tolerate frost and is best kept above 5^0C \ they form impenetrable barriers and are probably suited to living fences \ day flowering in summer \ the reddish-green, spherical fruit grows to 12mm diameter and is bitter, but edible.

Stenocereus eruca
Machaerocereus eruca

creeping devil

Baja in California and north west Mexico (vulnerable) \ low, columnar, shrub cactus \ stems to 3m long grow along the ground and takes root as they go \ does not tolerate frost and is best kept above 5^0C \ full sun needed \ nocturnal flowers in spring \ edible fruit \ plant crushed and thrown into water to stun fish \ grows from seed at 20-30^{0}C with 13-14 hours daylight per day.

Stenocereus fimbriatus

Lemaireocereus hystrix
Ritterocereus hystrix
Stenocereus hystrix
Cuba, Haiti, Dominican Republic, Jamaica, Puerto Rica \ edible fruit.

Stenocereus griseus

Stenocereus deficiens
Stenocereus eburneus
 Lemaireocereus deficiens
Lemaireocereus griseus
Ritterocereus deficiens
Ritterocereus griseus
Venezuela, Colombia, Trinidad, Tobago, Antilles, Mexico \ does not tolerate frost and is best kept above 5^0C \ cultivated in Mexico for the edible fruit \ stem part sliced and fried like potatoes \ the ash is used as a potassium fertiliser.

Stenocereus gummosus

Machaerocereus gummosus

pitahaya agria
creeping devil
dagger

North west Mexico \ low, scrambling, columnar, shrub

cactus to 1m tall and often with arching stems branching from the base \ does not tolerate frost and is best kept above 5^0C \ full sun on calcareous, rich soil \ night flowering in early summer \ round, red, edible fruit has a tart yet sweet flavour and is eaten fresh, dried and in unsweetened jams \ edible seed \ stem juice used to stupefy fish \ grows from seed at 20-30^{0}C with 13-14 hours daylight per day.

Stenocereus montanus

Lemaireocereus montanus
Mexico \ edible fruit.

Stenocereus pruinosus

Lemaireocereus pruinosus
Ritterocereus pruinosus
Mexico \ edible fruit.

Stenocereus queretaroensis

Lemaireocereus queretaroensis
Ritterocereus queretaroensis

pitahaya
Mexico \ cultivated for it's edible red fruit which has a sweet, delicious flavour \ typical fruit yield is 5-10 tonnes per hectare.

Stenocereus standleyi

Lemaireocereus standleyi
Ritterocereus standleyi
Mexico \ edible fruit.

Stenocereus stellatus

Lemaireocereus stellatus

joconostle
Mexico \ shrubby cactus \ edible spiny fruit is said to make the best of the cactus fruit jams (joconostle jam) which is eaten hot with white goat cheese \ stem contains alkaloids: Mescaline, 3,4-

Dimethoxyphenethylamine and 4-Hydroxy-3,5-dimethoxyphenethylamine.

Stenocereus thurberi

Cereus thurberi
Lemaireocereus thurberi
Marshallocereus thurberi

organ pipe cactus
pitahaya dulce

Warmer parts of the Sonoran Desert, in southern Arizona, Baja California, and Sonora, Mexico \ full sun \ day flowering in the summer \ long-lived, fast-growing to 7m \ several to many tall columns (200mm diameter) arising candelabra-like from the base or not far above it \ valued for the delicious, large, sweet fruit which is described as watermelon, fig, coconut flavoured and eaten fresh, dried, made into jams, jellies and wines \ seed eaten \ stems eaten \ usually bears twice a year \ petals of the funnel-shaped, white flowers are eaten \ seed pounded to a meal \ cuttings often planted closely in rows for hedges.

Stenocereus treleasii

USA, Mexico \ shrubby, columnar, weak-stemmed cactus \ best in full sun \ day flowering in summer \ edible fruit.

Stetsonia coryne
Cereus chacoanus

tooth pick

North west Argentina, Paraguay, Bolivia \ only species in this genus \ tree-like cactus to 8m tall with a trunk to 400mm thick \ 8-9 ribbed blue-green stems \ black spines as juvenile form which fade to white with black tips \ requires full sun in well drained soil \ hardy to minus 5^0C \ white, funnel-shaped, nocturnal flowers 80mm long form at night in summer \ edible fruits \ flesh contains mescaline, similar to San Pedro.

Strombocactus disciformis
Mammillaria discifomis
Echinocactus disciformis
Ariocarpus disciformis
Echinocactus turbiniformis
Strombocactus pulcherrimus
Ariocarpus pulcherrimus

peyotillo

Mexico \ only species in genus \ close resemblance to *Lophophora williamsii* \ narcotic or medicinal uses.

Turbinicarpus pseudomacrochele
Pelecyphora pseudopectinata
Normanbokea pseudopectinata

peyotillo

Mexico \ usually solitary or slowly clumping, up to 4 cm tall and 4 cm in diameter \ plant contains 2.48% mescaline.

Glossary

alkaloid: a class of basic nitrogenous organic compounds which occurs in plants. Many have medicinal or psycho-active effects.

awl: pointed instrument for piercing small holes in leather

cabeza: large stem of agave plants after leaves have been removed.

cathartic: an agent that works to empty the bowels (laxative).

CITES: The Conservation on International Trade in Endangered Species of Wild Flora and Fauna – an international organisation to attempt to protect exploitation of wild populations of endangered life forms.

cladodes: young stem segments of prickly pear also called joints, pads or nopales. Once sliced and prepared for eating they are generally called nopalitos.

digitalin: chemical, first found in digitalis, which is a powerful cardiac drug in small doses but deadly poisonous in large doses.

diuretic: agent for increasing the flow of urine.

earth oven: used by Amerindians of California to cook food \ a hole was dug in the ground, lined with rocks and heated by building a fire in it \ when the rock lining was thoroughly heated, the fire was raked out and food was placed in the hot pit and covered with leaves, hot rocks and earth \ cooking often took as long as 48 hours and some foods required a second fire to be built on top of the oven whilst the food cooked.

emetic: an agent that cause vomiting.

false peyote: agents which have similar effects to, but are not a *Lophophora sp.*

globose: globe shaped.

mescal: a loose term referring either to the peyote button or the distilled juice of fermented agave sap. Mescal bean may also refer to *Sophora secundifolia* which is poisonous but credited with hallucinogenic properties.

mescaline: alkaloid ($C_{11}H_{17}NO_3$) which causes hallucinations if ingested.

nopalitos: fresh or cooked young stems of opuntia which have been prepared for eating. Nopalitos consumption is recommended for those with diabetes, high cholesterol and prostrate problems.

provenances: plants of the same species can have very different characteristics by virtue of having adapted to a different location.

stem: may also be referred to as **pad** or **leaf.**

stomachic: an agent that stimulates digestion.

tuna: fruit of opuntia.

tiswin **water**:" a liquor made from fermented maize.

tuber: underground, succulent, roundish stem.

tubercles: small knob-like projections.

pepinillos: pickled opuntia stems.

pulque: the word was borrowed by the Spanish from the Araucanian Indians of Chile \ it is a drink made by making a cavity in an agave plant, by taking out the flowering stem and part of the main body of the plant, just as it starts to flower \ the sugary sap that runs out from the flower stem fills the cavity and ferments (4-8% alcohol).

raicilla: a form of mescal (spirit).

yucca shoot: flower stalk.

Appendix 1
Suppliers of Seeds and Plants

This list is by no means complete. There should be enough contacts listed to access most of the species in the book. The comments below are personal opinions. They are there to give the reader an idea of which company to approach for their needs. I did not make the assessments all at the same time so there is an element of subjective opinion.

I imagine that most people will be chasing established plants or small packs of seed.

USA

Abbey Garden Cactus
PO Box 2249
La Habra
CA 90632
Mail order specialists.
$2 for catalogue.

Aldridge Nursery, Inc
PO Box 1299
Von Ormy
TX 78073

Allies
PO Box 2422,
Sebastopol
CA 95473
Range of seed including some hallucinogenic cacti.

Aztekakti
PO Box 26126
El Paso
TX 79926
Range of seed of desert and sub-tropical seeds including cacti and AGAVACEAE, mostly in lots of 1000 or more.

Bach's Cactus Nursery
8602 North Thornydale Road
Tucson, AZ 85741

Bobtown Nursery
16212 Country Club Rd.
Melfa
VA 23410

Burks Nursery
PO Box 1207
Benton,
Arkansas 72015-1207

Cactus by Dodie
934 East Metler Rd
Lodi
CA 95242

Cactus Data Plants
9607 Ave S-12
Littlerock
CA 93543
 Seed and plants.

Cactus by Mueller
10411 Rosedale Hwy
Bakersfield, CA 93308

Cactus Unlimited
Gardena Drive
Cupertino, CA 95014

Carter Seeds
475 Mar Vista Drive
Vista
CA 92083
Request special cactus and succulent list.
WWW.carterseeds.com

Crump Greenhouse
Box 185
225 S. Pleasant Ave.
Buena Vista, Colorado 81211

Cycadia
17337 Chase St.
Northridge CA 91325

Deep Diversity
PO Box 15700
Santa Fe
NM 87506-5700

Desert Dan's Nursery Seed Company
Minotola, New Jersey 08341

Desert Enterprises
PO Box 23
Morristown
AZ 85342
Bulk seed (min. 1 pound) of 21 Cacti species listed in this book.

Desert Moon Nursery
PO Box 600
Veguita
NM 87062

Desert Nursery
1301 South Cooper
Deming
NM 88030-5028

Drummong Nursery and Greenhouse
Route 1, Long Road
De Soto,
Missouri 63020

EPI WORLD
10607 Glenview Ave.
Cuperntino
CA 95014
Specialists in epiphyllum cacti.

Exotic Seeds From Around The World
1814 NE Schuyler
Portland
OR 97212

Fernwood Plants
PO Box 268
Topanga CA 90290

Florida Cactus Inc.
PO Drawer D
Apopka, FL 32703

George W. Park Seed Co.
1 Parkton Ave.,
Greenwood
SC 29647-0001

Grigsby Cactus Gardens
2326 & 2354 Bella Vista
Vista CA 92083
Plants only.

Guy Wrinkle Exotic Plants
11610 Addison St.
North Hollywood
CA 91601
www.exotic-plants.com
Large selection of succulent plants.

Henrietta's Nursery
1345 North Brawley
Fresno CA 93711

High Country Gardens
2902 Rufina St.
Santa Fe
NM 87505-2929

Hillis Nursery Co. Inc.
92 Gardner Rd.
McMinnville
TN 37110

Hines Nurseries
PO Box 11208
Santa Ana
CA 92711

Homan Brother Seed
1540 Happy Valley Rd.
Phoenix
AZ 85027

Horus Botanicals
HCR 82
Box 29
Salem
AR 72576
CACTI seed limited to hallucinogens.

Huntington Botanic Gardens
1151 Oxford Rd.
San Marino
CA 91108
Seed and Plants.

Hurov's Seeds & Botanicals
PO Box 1596
Chula Vista
CA 91912

JL Hudson,
Seedsman
Star Route 2
Box 337
La Honda
CA 94020
Limited agave and cacti but extensive
gene types of *Trichocereus peruvianus*
and good general Ethnobotanical range.

Intermountain Cactus
1478 North 750 East
Kaysville
UT 84037

Joseph Brown Wild Seeds and Plants
7327 Hoefork Lane
Gloecester Point
VA 23062

K&L Cactus Nursery
9500 Brook Ranch Rd East
Ione
CA 95640-9417

Kimura International Inc.
18435 Rea Ave.
PO Box 327
Aromas CA 95004

Las Palitas Nursery
Las Palitas Rd.
Santa Margarita
CA 93453

Legendary Ethnobotanical Resorces
16245 S.W. 304th St.
Leisure City
FL 33033
www.ethnobotany.com
Large range but no agaves and few cacti.

Living Stones Nursery
2936 North Stone Ave.,
Tuscon
AZ 85705
Large selection of AGAVACEAE and
cacti plants.

Loehman's Cacti and Succulents
PO Box 871
Paramount
CA 90723

Logee's Greenhouses
141 North St.
Danielson
CT 06239

Louisiana Nursery
5853 Highway 182
Opelousas
LA 70570

McClure and Zimmerman
PO Box 368
Friesland
WI 53935

Mellinger's Inc.
2310 West South Range Rd.
North Lima
OH 44452-9731

Mesa Garden
PO Box 72
Belen,
NM 87002
***** Huge range of cacti and
AGAVACEAE seed from many different
provenances. Well priced packets of
seeds and also available in larger
amounts.

Missouri Wildflowers Nursery
9814 Pleasant Hill Rd.
Jefferson City,
MO 65109-9805

Mistletoe Quality Seeds
780 North Glen Annie Rd.
Goleta
CA 93117
Seed of some AGAVACEAE (Minimum
trade pack of 1 ounce or 1000 seeds.)

Neon Palm Nursery
3525 Stony Point Rd.
Santa Rosa
CA 95407

New Mexico Cactus Research
PO Box 787
Belen
NM 87002

Oak Hill Gardens
PO Box 25
Dundee
IL 60118-0025

Oakhill Gardens
1960 B Cherry Knoll Rd.
Dallas Oregon 97330

Oregon Exotics Nursery
1065 Messinger Rd.
Grants Pass
OR 97527
Exciting selection of edible plants and
seed, including a few fruiting cacti.

PJT Botanicals
PO Box 49
Bridgewater
MA 02324-1630
Selection of cacti seed and some plants,
catalog $3 in USA, $5 elsewhere.

Plants of the Southwest
Agus Fria,
Rt. 6
Box 11A
Santa Fe,
NM 87501
www.plantsofthesouthwest.com
Seeds and plants of a few species of the
family AGAVACEA.

Rainbow Gardens Nursery and Bookshop
1444 East Taylor St.
Vista
CA 92084

Redlo Cacti Inc.
2315 NW Circle Blvd
Corvallis, Oregon 97330

Serra Gardens
3314 Serra Road
Malibu
CA 90265
HTTP//WWW.iquest.net/serra/index.htm
l

Sheffield's Seed Co.
273 Auburn Rd.
Route 34
Locke
NY 13092
http://www.sheffields.com
Some AGAVACEAE bulk seed only

Shein's Cactus
3360 Drew St.
Marina
CA 93933
Cacti plants on offer.

Singer's Growing Things
17806 Plummer St.
Northridge, CA 91325

Sourcepoint Organic Seeds
1647 2725 Rd.
Cedaredge
CO 81413

Southwestern Native Seeds
Box 50503
Tucson
AZ 85703
Selection of AGAVACEAE seed.

Springbrook Gardens Inc.
PO Box 388
Mentor
OH 44061

Sunlight Gardens
174 Golden Lane
Andersonville
TN 37705

Sunny Lands Seeds
Box 385
Paradox
CO 81429

Sunshine Farm and Gardens
Route 51D
Renwick
WV 24966

Tanque Verde Greenhouses
10810 East Tanque Verde Rd.
Tucson, AZ 85749

The Banana Tree
715 Northampton St.
Easton
PA 18042
http://www.banana-tree.com

The Cactus Farm
Route 5
Box 1610
Nacogdoches
TX 75961

The Flowery Branch Seed Co.
PO Box 1330

Flowery Branch
GA 30542

The Plumeria People
910 Leander Dr.
Leander
TX 78614

The Primrose Path
R.D. 2, Box 110
Scottdale
PA 15683

The Theodore Payne Foundation
for Wildflowers and Native Plants
10459 Tuxford St.
Sun Valley
CA 91352-2126
Seed packs or bulk of a few
AGAVACEAE and *Opuntia basilaris*.

Tripple Brook Farm
37 Middle Rd.
Southampton
MA 01073

Tropical Fruit Trees
7341 121st Terrace North
Largo
FL 33773

Van Bourgondien Bros.
PO Box 1000
Babylon
NY 11702-9004

Wayside Gardens
1 Garden Lane
Hodges
SC 29695-0001

Wildflower Nursery
1680 Highway,
25-70
Marshall,
NC 28753

World Wide Exotic Seed Co.
PO Box 57770
Webster
TX 77598

Europe

B&T World Seeds,
sari – Paguignan,
Olonzac 34210
FRANCE
http://www.b-and-t-world-seeds.com
Huge range, high price.

Chiltern Seeds
Bortree Stile
Ulverston
Cunbria LA12 7PB
ENGLAND
Small range of cacti and AGAVACEAE
seed, beware of packs of mixed species,
fairly costly.

Doug and Vivi Rowland
200 Spring Road
Kempston, Beds
MK42 8ND
ENGLAND
Good range of cacti and AGAVACEAE
seed in small or large quantities at
reasonable prices.

Gerhard Köhres
Wingerstrasse 33,
Erhausen/Darmstadt
D-64387
GERMANY
Good range of AGAVACEAE and cacti
seed in small or large quantities at
reasonable price.

Roy Young Seeds
23 Westland Chase
West Winch
King's Lynn
Norfolk
PE33 0QH
ENGLAND
Good selection of cacti seed (min.
amount 500 seed).

Southwest Seeds Ltd.
200 Spring Road
Kempston, Bedford
MK42 8ND
England

Australia

Arizona Cacti Nursery
RMB 111 Windsor Rd.
Box Hill
NSW 2765
(Wholesale generic agave and cacti
plants.)

Australian Cactus and Succulent Supplies
Lot 3 Cessnock Rd.
Sunshine
NSW 2263

Buena Vista Nursery
Wynyard Ave.
Rossmore
NSW 2171

Cacti Collectors Corner
Garden World Nursery
Spring Vale Rd.
Keysborough Vic 3173

Cactus Garage
3 Edward St.
Bayswater
Vic. 3153
Mail order cacti plants.

Ellison Horticultural Pty. Ltd.
PO Box 365
Nowra
NSW 2541

Hamilton's World of Cacti
Lot 2 4th Ave.
Llandilo
NSW 2760

John Spencer
1 Little Addison St.
Goulburn
NSW 2580

ML Farrar Pty. Ltd.
PO Box 1046
Bomaderry
NSW 2541

Mrs Joscelyn Burnett
'Andoran' Drakes Forest
NSW 2508

Orana and Mexicana Cacti and Succulent
Gardens
57 Wamboin St.
Gilgandra NSW 2827

R & C Metcalfe
Grittenden Road
Glasshouse Mountains
Q 4518

R Field
'Whiroa'
Tennyson Vic 3572

Sunshine Coast Cactus Nursery
5 Daniel St.
Nambour Q 4560

Tarrington Exotics
Box 40
Teesdale Vic 3328
www.tarex.com.au
Large range of succulent plants.

Thornwood Gardens
PO Box 282
Strathalbyn
SA 5255
Mail order cacti plants.

New Zealand

Planters World Garden Centre
High St.
Lower Hutt

Oderings Nursery Christchurch Ltd.
92 Stourbridge St.
Spreydon, Christchurch

Peter B. Dow & Co.
PO Box 696,
Gisborne 3800

Others

Israflora
PO Box 502,
Kiryat-Bialik 27093
ISRAEL
Sell forest seed of arid areas, list *Yucca
aloifolia*.

Kumar International
Ajitmal 206121
Etawah
Uttar Pradesh 244 715
INDIA
Agave americana and *Pereskia
grandifolia* seed.

Appendix 2 Cactus and Succulent Societies

USA

Cactus and Succulent Society of America
Box 3010
Santa Barbara
CA 93105 USA

Austin Cactus and Succulent Society
602 Tce. Mt Dve.
Austin
TX 78746

Cactus and Succulent Society of
California
547 Ealanor Pl.
Hayward
CA 94544

Cactus and Succulent Society of Greater
Chicago
7706 N Harlem
Niles
ILL 60648

Cactus and Succulent Society of Hawaii
1218 Akupa Pl.
Kailua
HI 96734

Cactus and Succulent Society of
Maryland
731 Brook Wood Rd.
Baltimore
MD 21229

Cactus and Succulent Society of New
Jersey
363 Washington Ave.
Rutherford
NJ 07070

Cactus and Succulent Society of San Jose
1895 Crinan Dve.
San Jose
CA 95112

Cactus and Succulent Society of Tulsa
139 SE Fenway Pl.
Bartlesville
OK 74006

California Cactus Growers Inc.
1860 Monte Vista Dve.
Vista
CA 92083

Carmichael Cactus and Succulent Society
3808 French Ave.
Sacramento
CA 95821

Cascade Cactus and Succulent Society
2615 NE 137th St.
Seattle
WA 98125

Central Arizona Cactus and Succulent
Society
13493 N 88th Pl.
Scottsdale
AZ 85260

Central Arkansas Cactus and Succulent
Society
3500 Bowman Rd.
Little Rock
AR 72211

Central Ohio Cactus and Succulent
Society
1446 W 2nd Ave.
Columbus
OH 43212
USA

Central Oklahoma Cactus and Succulent
Society
11201 Draper Ave.
Choctow
OK 73020

Cincinnati Cactus and Succulent Society
6501 Hamilton Ave.
Cincinnati
OH 45224

Coastal Bend Cactus and Succulent
Society
1316 Cambridge Dve.
Corpus Christi
TX 78415

Connecticut Cactus and Succulent
Society
RFD #1 Bx 86 Roger Ft
Lebanon
CT 06249

Cox Arboretum Cactus and Succulent
Society
5438 Camelia Pl.
Dayton
OH 45429

Fresno Cactus and Succulent Society
3015 Timmy St.
Clovis
CA 93612

Fort Worth Cactus and Succulent Society
3403 NW 25th St.
Fort Worth TX

Gates Cactus and Succulent Society
307 Westwood Lane
Redlands
CA 92373

Henry Shaw Cactus Society
4050 Shenandoah St.
St. Louis
MO 63110

Houston Cactus and Succulent Society
8822 Bobwhite Dve.
Houston
TX 77074

Indiana Cactus and Succulent Society
4380 W 1200 N-90 St.
Roanoke
IN 46783

Kansas City Cactus and Succulent
Society
7109 Riggs St.
Overland Park
KS 66204

Las Vegas Cactus and Succulent Society
3814 Pecan Lane
Las Vegas
NV 89115

Lincoln Cactus and Succulent Society
2755 Pear St.
Lincoln, NE

Long Beach Cactus and Succulent
Society
1303 Broad Ave.
Wilmington
CA 90744

Long Island Cactus and Succulent Society
5046 Clearview Expwy
New York
NY 11364

Los Angeles Cactus and Succulent
Society
21129 Merridy St.
Chatsworth
CA 91311

Louisiana Cactus and Succulent Society
106 Aster Lane
Waggaman
LA 70094

Michigan Cactus and Succulent Society
3921 Auburn Dve.
Royal Oak
MI 48072

Monterey Bay Area Cactus and Succulent
Society
134 Trabing Rd.
Watsonville
CA 95076

National Capital Cactus and Succulent
Society
1619 Millersville Rd.
Millersville
MD 21108

New Mexico Cactus and Succulent
Society
423 13th NW St.
Albuquerque
NM 87102

North Texas Cactus and Succulent
Society
4128 Hackmore Loop
Irving
TX 75061

Omaha Cactus and Succulent Society
12618 Orchard St.
Omaha
NE 68137

Palomar Cactus and Succulent Society
2711 Athens Ave.
Carlsbad
CA 92008

Peninsula Succulent Society
848 Miramar Tce.
Bohmont
CA 94002

Philadelphia Cactus and Succulent
Society
83 Belair Rd.
Warminster
PA 18974

Redwood Empire Cactus and Succulent
Society
350 4th St.
Lakeport
CA 95453

Sacramento Cactus and Succulent Society
3836 65th St.
Sacramento
CA 95820

San Antonio Cactus and Xerophyte
Society
230 Killarney Dve.
San Antonio
TX 78223

San Gabriel Cactus and Succulent
Society
14731 La Forge St.
Whittier
CA 90603

South Bay Epiphyllum Society
1801 W 27th St.
San Pedro
CA 90732

Stockton Cactus and Succulent Society
519 Milo Rd.
Modesto
CA 95350

Texas Association of Cactus and
Succulent Society
11807 Dover St.
Houston
TX 77031

Wasatch Cactus and Succulent Society
3234 S 800 East St.
Salt Lake City
UT 84106

Canada

Desert Plant Society of Vancouver
456 E 45th Ave.
Vancouver
BC V5W IX4

Toronto Cactus and Succulent Club
RR2 Georgetown
Ont L7G 4S5

Europe

National Cactus and Succulent Society
The Herbarium
Royal Botanic Gardens, Kew
Richmond, Surrey
TW9 3EA
England

Kakteen Und Andre Sukkulenten
Ahornweg 9
D-7820 Titisee
Neustadt
West Germany

Deutsche Kakteen
Moorkamp
22D-3008 Garbsen 5
West Germany

Australia

Australian Society of Cacti and
Succulents
53 Manningtree Rd
Hawthorn, Vic, 3122

Succulent Publications of South Australia
Inc.
PO Box 572
Gawler, SA 5118

Australian Cactus and Succulent
Association
Lyn Schultz
Teesdale, Vic 3328

Cactus and Succulent Society of the ACT
GPO Box 2499
Clive, ACT 2600

Cactus and Succulent Society of NSW
PO Box 36
Woolahara, NSW 2025

Central Coast Cactus and Succulent Club
of NSW
10 Blackford Ave.
Kanwal, NSW 2559

Epiphytic Cacti-Asclepiadaceae Society
of Australia
32 Wynyard St.
Rossmore, NSW 2171

Goulburn Cactus and Succulent Club of
NSW
1 Little Addison St.
Goulburn, NSW 2580

Murray Region Cactus and Succulent
Club
102 Thomas Mitchell Dve.
Wodonga, Vic 3690

Summerland Cactus and Succulent
Society
37 Laurel Ave.
Casino, NSW 24570

Western Suburbs Cactus Club
16 Paterson Crescent
Fairfield, NSW 2165

Central Queensland Succulent Society
PO Box 6407
North Rockhampton, Q 4006

Far North Queensland Succulent Society
17 Suhle St.
Edmonton, Q 4869

Cactus and Succulent Society of South
Australia
PO Box 37
Rundle Mall, SA 5000

Cactus and Succulent Society of SA, SE
Branch
Attamurra Cottage
Attamurra Rd.
Mount Gambier, SA 5290

Southern Tasmania Cactus and Succulent
Club
1 Willow Walk
Austin's Ferry, Tas 7011

Ballarat Cactus and Succulent Society
8 Margaret St.
Wendouree, Vic 3355

Cactus and Succulent Society of Australia
19 Kingsley St.
Camberwell, Vic 3124

Geelong Cactus and Succulent Club
c/- Mr J Edmonds
Torquay Rd.
Mount Duneed, Vic 3126

Sunrasia Cactus and Succulent Society
PO Box 1475
Mildura, Vic 3500

Bunbury Cactus and Succulent Study
Group
Lot 100 Bridge St.
Boyanup, WA 6105

Cactus and Succulent Society of Western
Australia
348 Hardy Rd.
Cloverdale, WA 6105

Epiphytic Cacti and Hoya Society of
Australia
PO Box 210
Morley, WA 6062

WA Cactus and Other Succulent Study
Group
17 Parramata Rd.
Doubleview, WA 6018

New Zealand

New Zealand Cactus and Succulent
Society
164 Massey St.
Frankton, Hamilton
New Zealand

	Opuntia sp.	*Hylocereus undatus*	*Pereskia aculeata* fruit	*Pereskia aculeata* leaves
Moisture	80.1%	83%	91.4%	
Protein	1.0%	0.16 – 0.23%	1.0%	
Fat	1.8%	0.21 – 0.61%	0.7%	6.8 – 11.7%
Carbohydrates	13.1%		6.3%	
Crude Fibre	4.2%	0.7 – 0.9%	0.7%	9.1 – 9.6%
Ash		0.54 – 0.68%	0.6%	20.1 – 21.7%
Calcium mg/100g	26	0.63 – 0.88%	174	2.8 – 3.4%
Phosphorous mg/100g	29	30.2 – 36.1	26	
Iron mg/100g	0.3	0.55 – 0.65	trace	
Magnesium				1.2 –1.5%
Carotene mg/100g		0.005 – 0.012	3215 I.U.	
Thiamin mg/100g		0.028 – 0.043	0.03	
Riboflavin mg/100g		0.043 – 0.045	0.03	
Niacin mg/100g		0.297 – 0.430	0.9	
Ascorbic Acid mg/100g	15	8.0 –9.0	2	

This Dracaena draco is 140 years old.
A storm caused it to become unstable
so metal props were employed to preserve it.
Brisbane Botanical Gardens.

References

Altshul, S. Drugs and Foods from Little Known Plants. (Harvard University 1973)

Balls, E. K. Early Uses of California Plants. (University of California Press 1965)

Basehart, H. W. Apache Indians XII. Mescalero Apache Subsistence, Patterns and Socio-political Organisation. (Garland Publishing 1974)

Bell, W. H. and Castetter E. F. Ethnobiological Studies in the American Southwest VII. The Utilisation of Yucca, Sotol and Beargrass by the Aborigines in the American Southwest. (University of New Mexico 1941)

Bremnes, L Herbs. (Harper Collins 1994)

Britton, N. and Rose, J. The Cactaceae, Vols I to IV. (Dover)

Burskirk, W The Western Apache: Living With the Land Before 1950. (University of Okalahoma Press 1986)

Castetter, E. F. Ethnobiological Studies in the American Southwest. (University of New Mexico 1935)

CERES, FAO Review on Agriculture and Development. (FAO 1976-96)

Chapman, P. and Martin, M. An Illustrated Guide to Cacti and Succulents. (Salamander Books - undated)

Clark, D. E. (editor) Desert Gardening. (Sunset Books 1973)

Clodsley-Thompson, The Desert. (Orbis Publishing 1977)

Dastur, J. F. Medicinal Plants of India. (Taraporevala, Sons & Co. 1977)

Dastur, J. F. Useful Plants of India. (Taraporevala, Sons & Co. 1977)

Davis, J. B., Kay, D. E. and Clark, V Plants tolerant of arid, or semi-arid conditions with non-food constituents of potential use. (Tropical Products Institute 1983)

Curtin, L. S. M. By the Prophet of the Earth. (San Vincente Foundation 1947)

Dawson, E. Some Ethnobotanical Notes on the Seri Indians. (Desert Plant Life 1944)

Duffield M. R. and Jones, W. D. Plants for Dry Climates. (HP Books 1981)

Duke, J. A. Dr. Duke's Phytochemical and Ethnobotanical Databases. (US Department of Agriculture. http://www.ars-grin.gov/cgi-bin/duke/ethnobot.pl)

Elmore, F. H. Ethnobotany of the Navajo. (New Mexico School of American Research 1944)

Facciola, S. Cornucopia - A Sourcebook of Edible Plants. (Kampong 1990)

Facciola, S. Cornucopia II - A Sourcebook of Edible Plants. (Kampong 1998)

Greives, M. A Modern Herbal. (Penguin 1980)

Grubber, H. Growing the Hallucinogens. (High Times/Level Press 1973)

Hendrick, U. P. (ed.) Sturtevant's Edible Plants of the World. (Dover 1972)

Howes. P.G. The Cactus Forest and it's World. (Little, Brown and Co. 1954)

Hrdlicka, A. Physiological and Medical Observations Among Indians of SW United States and Northern Mexico. (SI-BAE Bulletin #34)

Hudson, J. L. Ethnobotanical Catalogue of Seeds. (Star Route 2, Box 337 La Honda, California, 94020 USA. various editions)

Hunt, D. CITES Cactaceae Checklist. (Royal Botanic Gardens Kew 1992)

Innes, C. & Glass, C. The Illustrated Encyclopedia of Cacti. (Knickerbocker Press 1997)

Indian Council of Agricultural Research Handbook of Agriculture. (1987)

Jacobs, D. Pitaya (Hylocereus undatus), a potential crop for Australia. (West Australian Nut and Tree Crops Association Yearbook 1998)

Johns, L. and Stevenson, V. The Complete Book of Fruit. (Angus and Robertson 1979)

Jolly, A. A World Like Our Own: Man and Nature in Madagascar. (Yale 1980)

Kunkel, G. Plants for Human Consumption. (Koeltz Scientific Books 1984)

Lamb, B. M. A Guide to the Cacti of the World. (Angus and Robertson 1991)

Leonard, J. N. Latin American Cooking. (Time-Life Books 1970)

Limón, E.M. Tequila, the Spirit of Mexico. (Abbeville Press 2000)

Lust, J. The Herb Book. (Bantam Books 1980)

MacMillan, H. F. Tropical Planting and Gardening, with special reference to Ceylon. (MacMillan and Co. 1952)

Mann, J. Cacti Naturalised in Australia and Their Control. (Dept. of Lands, Queensland. 1970)

Martin, F. et al ECHO's Inventory of Tropical Vegetables. (ECHO undated)

Meitzner, L. & Price, M. Amaranth to Zai Holes \ Ideas for Growing Food Under Difficult Conditions. (ECHO)

Menninger, E. A. Fantastic Trees. (Horticultural Books 1975)

Mollison, B. The Permaculture Book of Ferment and Human Nutrition. (Tagari 1993)

Morton, J. F. Fruits of Warm Climates. (Julia F Morton 1987)

Nobel, Park, S. Environmental Biology of Agaves and Cacti. (Press Syndicate of University of Cambridge 1988)

Nobel, Park, S. Remarkable Agaves and Cacti. (Oxford University Press 1994)

Palmer, G. Shuswap Indian Ethnobotany. (Syesis 1975)

Pilbeam, J. Cacti for the Connoisseur. (Batsford Ltd. 1998)

Robbins, W. W. and Harrington, J. P. Ethnobotany of the Tewa Indians. (Smithsonian Institution 1916)

Rowley, G. The Illustrated Encyclopedia of Succulents. (Salamander Books 1978)

Saunders, C. F. Edible and Useful Wild of The United States and Canada. (Dover 1976)

Schuster, D. What Cactus is That? (Pierson & Co.1990)

Slay, R. Useful Cactus, Agave and Aloe Species. (Tagari undated)

Smith, M.S. Narcotic and Hallucinogenic Cacti of the New World. (better Days Publishing 2000)

Stafford, P. Psychedelics Encyclopedia. (Tarcher, 1983)

Swank, G. R. The Ethnobotany of the Acoma and Laguna Indians. (University of New Mexico 1932)

Swinbourne, R. F. G. Sansevieria in Cultivation in Australia (Adelaide Botanic Gardens 1979)

Tate, J. L. Cactus Cookbook. (Cactus and Succulent Society of America 1990)

Twentieth Century Alchemist Legal Highs. (1973)

US Dept. of Agriculture Ethnobotany Database. (http://probe.nal.usda.gov:8300/index.html)

van Epenhuijsen, C. W. Growing Native Vegetables in Nigeria. (FAO 1974)

Vestal, P. A. Ethnobotany of the Ramah Navaho. (Harvard University 1952)

Weiner, M. A. Earth Medicine, Earth Food. (MacMillan Publishing 1980)

Wessels, A. B. Spineless Prickly Pears. (Perskor Publishers 1988)

White, A. Herbs of Ecuador: medicinal plants. (Ediciones Libri Mundi 1982)

White, L. A. Notes of the Ethnobotany of the Keres. (Papers of the Michigan Academy of Arts 1945)

Whiting, A. F. Ethnobotany of the Hopi. (Flagstaff 1939)

Wickens, G. E. and Goodwin, J. R. Kew International Conference on Economic Plants for Arid Lands. (George Allen and Unwin 1985)

Wilson, B A Manual of California Native Plants. (The catalogue of Las Palitas Nursery 1993)

Wolverton, B. C. Eco-Friendly House Plants. (Phoenix Illustrated 1997)

Young, H. (translator) Macdonald Encyclopedia of Trees. (Macdonald & Co. 1982)

Zeven, A. C. and de Wet, J. M. J. Dictionary of Cultivated Plants and their Regions of Diversity. (Centre for Agricultural Publishing and Documentation 1982)

http://www.greendealer-exotic-seeds.com/seeds/MedicinalHerbs.html

For latest updates from SARI
visit
www.permacultureplants.net

Index